GREEN RESEARCH, DEVELOPMENTS, AND PROGRAMS

IN SEARCH OF THE SILVER BULLET

ASSESSMENT OF ALTERNATIVES FOR TRICHLOROETHYLENE IN CLEANING OPERATIONS

Green Research, Developments, and Programs

Additional books in this series can be found on Nova's website under the Series tab.

Additional e-books in this series can be found on Nova's website under the e-book tab.

GREEN RESEARCH, DEVELOPMENTS, AND PROGRAMS

IN SEARCH OF THE SILVER BULLET

ASSESSMENT OF ALTERNATIVES FOR TRICHLOROETHYLENE IN CLEANING OPERATIONS

JASON P. MARSHALL
AND
HEIDI A. WILCOX

nova publishers
New York

Library of Congress Cataloging-in-Publication Data

ISBN: 978-1-63483-424-7
Library of Congress Control Number: 2015945690

Published by Nova Science Publishers, Inc. † New York

CONTENTS

List of Tables		**vii**
List of Figures		**ix**
Preface		**xi**
Chapter 1	In Search of the Silver Bullet: Assessment of Alternatives for Trichloroethylene in Cleaning Operations	**1**
Chapter 2	Background on Cleaning	**5**
Chapter 3	Why TCE	**11**
Chapter 4	Why Not TCE	**19**
Chapter 5	If Not TCE, Then What	**25**
Chapter 6	If No Silver Bullet, Then How	**33**
Chapter 7	If No Silver Bullet, Then How to Test	**57**
Chapter 8	If No Silver Bullet What to Test	**67**
Chapter 9	If No Silver Bullet Where Else to Look	**77**
Chapter 10	If No Silver Bullet Where To Run To, Who Else To Turn To	**97**
Chapter 11	Obstacles, Barriers and Issues, Oh My!	**111**
Chapter 12	Case Studies Outlining the Steps Taken to Replace TCE	**123**
Chapter 13	Future/Emerging Solvents/Next Steps	**147**

Chapter 14	Summary: Ending the Search for the Silver Bullet	**157**
About the Authors		**167**
Index		**169**

List of Tables

Table 3-1. Company Types Using TCE 12
Table 3-2. Contaminant Cleaned Using TCE 12
Table 3-3. Industry Classifications Potentially Using TCE 13
Table 3-4. Physical Properties of TCE 15
Table 4-1. TCE Regulatory Information 20
Table 4-2. TCE Emission Sources and Corresponding Total Annual Air Emissions of TCE as reported in the 2008 NEI 21
Table 4-3. Summary of TCE Health Effects 23
Table 5-1. Reported TCE Data in Massachusetts 26
Table 5-2. Massachusetts Companies Reporting More than 10 Years on TCE Usage 27
Table 5-3. Example Standardized Cleaning Protocols 28
Table 5-4. Questions to ask about cleaning processes 30
Table 6-1. Solvent Substitution Tools Evaluated in 2007 34
Table 6-2. P2OASys Criteria 37
Table 6-3. Green Screen Criteria 40
Table 6-4. CleanerSolutions Safety Screening 41
Table 6-5. Survey of Tools 43
Table 8-1. TCE Safety Screening Score 68
Table 8-2. Product Classifications for TCE Alternatives Tested at TURI Cleaning Lab 68
Table 8-3. Products Evaluated for TCE Replacement 69
Table 8-4. Top Cleaning Product Comparison 72
Table 8-5. Drop-In Solvent Comparison 73
Table 8-6. P2OASys Evaluation of Alternatives to TCE 74
Table 9-1. Safer Choice Program Master Criteria 80

Table 9-2. Green Seal GS 34 Criteria **86**
Table 9-3. Ecologo EHS Criteria **88**
Table 9-4. CleanGredients EHS Criteria for Solvents and Surfactants **90**
Table 9-5. Azeotrope Mixtures **94**
Table 9-6. Azeotrope Testing Results **94**
Table 10.1. Pollution Prevention Resources by EPA Regions **104**
Table 11.1. Solvent Replacement Projects Comparison **118**
Table 12.1. Soils Used for Tube Manufacturing **125**
Table 12.2. Safety Screen - Comparison of Potential Products **126**
Table 12.3. Soil 1 Removal Results **127**
Table 12.4. Non-Aqueous Products Selected **127**
Table 12.5. Soil 1 Removal Rates **128**
Table 12.6. Soil Removal Summary **128**
Table 12.7. Final EHS Comparison (P2OASys) **129**
Table 12.8. Safety Screen - Comparison of Potential Products **132**
Table 12.9. Soil Removal for All Soils **133**
Table 12.10. Added Solvents Safety and Performance **133**
Table 12.11. Vapor Degreasing with Ensolv **133**
Table 12.12. EHS Comparison of Selected Solvent **134**
Table 13.1. States Reporting TCE Disposal or Other Releases for 2013 **150**
Table 13.2. Top Industries Reporting TCE Disposal in 2013 **151**
Table 13.3. Pennsylvania Targeted Areas Using TCE **152**
Table 13.4. Kansas Targeted Areas Using TCE **152**
Table 13.5. Minnesota Targeted Areas Using TCE **152**
Table 13.6. Targeted Areas Using TCE **153**

LIST OF FIGURES

Figure 3-1. Vapor Degreasing Unit with Spray and Ultrasonic **14**
Figure 5-1. TCE Reduction in Massachusetts **27**
Figure 6-1. Search Method Refined **45**
Figure 6-2. Search Results **46**
Figure 6-3. Product Information Page **48**
Figure 6-4. Safety Screening Interface **49**
Figure 6-5. Product Comparisons **51**
Figure 6-6. Project Information **52**
Figure 6-7. Trial Details **53**
Figure 7-1 Phase I Cleaning Process **58**
Figure 7-2. Phase II Cleaning **60**
Figure 7-3. Phase III Cleaning **61**
Figure 7-4. Phase IV Part to be Cleaned **63**
Figure 7-5. Phase V Pilot Testing **64**
Figure 12-1. Tubes to be Cleaned **125**
Figure 12-2. Mill Building Location **141**
Figure 12-3. Interior Building Layout **141**
Figure 12-4. Facility located adjacent to Residential Housing **142**
Figure 12-5. Degreasing Equipment **142**
Figure 12-6. Open Top Degreaser **143**

PREFACE

When the Toxics Use Reduction Institute created its surface cleaning laboratory in the fall of 1993, they envisioned service that would provide support to companies in Massachusetts needing help replacing the chlorinated solvents being used for cleaning. The vision was a short term project lasting less than a decade. The project never quite turned out like they thought. Everyone thought a simple solution to solvent cleaning would appear, be adopted and the plight of TCE in cleaning would be wrapped up. As we all know, that was anything but the case. Early alternatives using water didn't fit the bill as many did not have as robust cleaning capabilities as TCE. Those companies that tried or did switch were left jaded on all aqueous cleaning for decades to come.

It was always troubling for us in the lab to hear other pollution prevention experts proclaiming the end of TCE in cleaning and watching them move on to other areas of interest. Not to be ignored, the TURI Lab has continued well beyond its original term limit and has moved into its third decade of helping companies find that silver bullet for replacing TCE in cleaning.

The truth about the TCE silver bullet is that there probably will never be one. No simple fix can be possible with the diverse nature of how TCE can be used. Every project we have seen is unique and just because one alternative worked for a particular case, it typically doesn't meet the exact needs of the next. To that end, the TURI Lab has refined and customizes its process to fit the needs of each project that comes into the lab. The silver lining to having no silver bullet is that this process works to identify several possible alternatives and works closely with the company in question to ensure success.

To help document what we have learned and how we go about solvent substitution in cleaning applications, we have compiled this book. We hope that the steps layout here will help others still struggling with the "simple" replacement of TCE or any other ineffective, unhealthy or unsafe cleaning product.

We would like to thank the past, current and future Cleaning Lab Staff. Without those that came before us, we would have had an even steeper hill to climb: John Bulko, Jay Jankauskas, Don Garlotto, Andy Bray, Tim Weil and most importantly Carole LeBlanc.

A thank you to the students who have allowed the lab to complete so much more than trying to do it alone. Hold on to your hats, this is a long list:

Francisco Abreu-Curiel, Parvathy Vijayan Babu, Shweta Bansal, Eyob Befekadu, Rhoda Bianchi, John Brunelle, Andrea Cheeseman, Junhee Cho, Samantha Couture, Shawn Crawford, Russell Curtis, Chris Damp, Suman Das, Purav Dave, Digvijay Devkota, Carla De La Cruz, Smith Gadd, Anni Geng, Abigail Giarrosso, Joshua Guerra, Alex Gyebi, Andy Harris, Dave Hout, Uyen Huynh (Jenny), Samuel Kibirige, Ravi Krishnappa, Nicholas Lanberg, Olantuji Lawal, George Liang, Binh Le, Johnny Le, Mina Le, Allison Leader, Todd MacFadden, Ephraim Massawe, Alicia Melvin, Scott Nadolna, Lam Nguyen, Loc Nguyen, Jonathan Oljey, Johanna Oviedo, Khoa Pham, Daniel Pina, Tony Pu, Ashwitha Rajagopal, Suterhman Ramesh, Luis Raudales, Natalie Regis, Cora Roelofs, Shyam Sarda, Mahima Tank, Jason Tarpley, Ruth Tejada, Kathleen Tenaglia, Vinh Tran, Prashant Triveidi, John Truong, Holly Thompson, Sai Rahul Tummala, Nicole Vayo, Titpisot Vorn (Nita), Chang Won, Lily Xu, and PaulYan.

Hopefully we didn't miss anyone. And for those still to come, we will need them to carry on the search for that possible silver bullet that is yet to be invented and or discovered.

Thanks to Linda Darveau at EPA Region 1 for her technical review of the material and her collaboration over the years. A special shout out to our sister agency the Massachusetts Office of Technical Assistance for their efforts in reducing hazardous solvents. Thanks to Dr. Fred Youngs for providing a final reality check of the final book. And a special thank you to Ashley Sanderson for her editing of the book. Being scientists, grammar is far from being our strongest skill set.

Chapter 1

IN SEARCH OF THE SILVER BULLET: ASSESSMENT OF ALTERNATIVES FOR TRICHLOROETHYLENE IN CLEANING OPERATIONS

ABSTRACT

There is no silver bullet solution for trichloroethylene in cleaning applications. Through many years of providing assistance to companies looking to replace TCE, the Toxics Use Reduction Institute Cleaning laboratory has developed a systematic way to provide process specific help in the quest for replacing TCE.

SILVER BULLET

A "silver bullet" has been defined as a quick solution to a difficult problem, a simple remedy for a difficult or intractable problem, a complete solution to a large problem, and a solution that seems magical [1, 2, 3]. Others use the term "silver bullet" to refer to an action which cuts through complexity and provides an immediate solution to a problem [4].

Using the term as the metaphor for a straightforward solution perceived to have extreme effectiveness fits nicely for describing the search for a universal drop-in solvent to replace trichloroethylene (TCE). The expectation that some new technological development or practice will easily solve this major prevailing problem is wishful thinking to say the least. So what options are out there beyond the TCE silver bullet? How does one solve the TCE conundrum without a simple remedy?

In the absence of this elusive silver bullet solvent, a systematic approach will need to be utilized to aid in the search for a replacement, drop-in or process change. The alternatives assessment process must include a way to compare the current solvent with its potential solvent or process change successor. Finding and adopting safer alternatives to toxic chemicals is a core aspect of what the Toxics Use Reduction Institute (TURI) Cleaning Lab does. For cleaning operations, TURI's Cleaning Laboratory has been working diligently for more than twenty years on assessing alternatives to TCE and other solvents. So far, no silver bullet has been found, but perhaps there is a silver lining in how the lab conducts its assessment process.

SILVER LINING - TURI CLEANING LAB

The passage of the Toxics Use Reduction Act (TURA) in 1989 by the Massachusetts legislature marked the creation of the Toxics Use Reduction Institute (TURI) at the University of Massachusetts Lowell. TURI collaborates with businesses, community organizations, and government agencies to reduce the use of toxic chemicals, protect public health and the environment, and increase the competitiveness of Massachusetts businesses.

TURI programs help companies and communities reduce the use of toxic chemicals. These program areas include training, business working groups, policy research, library, grants, and a cleaning laboratory.

Training at TURI provides education and workshops that offer tools and technical information needed to reduce toxics at a company's facilities. The main program includes a 40 hour course that trains professionals to be become certified Toxics Use Reduction Planners. The business working group program focuses on peer mentoring and other collaborative projects to help businesses meet environmental challenges. TURI develops policy analysis to support regulatory decision making with the TURA program and provides information on state and national toxics policy issues. The TURI library is part of the Massachusetts library system with a focus on information related to toxics use reduction. The Institute provides grant funding for industry, community groups, and university researchers. Along the research lines, the Institute operates a laboratory for safer cleaning.

The Cleaning Laboratory at UMass Lowell (Cleaning Lab) tests the performance of both industrial and janitorial cleaning products to encourage companies, institutions, and product formulators to choose and develop safer substitutes. The Cleaning Lab was established in late 1993 in response to the U.S. Environmental Protection Agency's Toxic Release Inventory (TRI), which revealed that many of the state's industries such as metalworking and electronics use several hazardous chemicals recognized for their excellent solvating powers [5].

The Cleaning Lab focuses on industrial parts cleaning to support Massachusetts and manufacturers in other states in using safer alternatives to existing toxic chemical solutions. The Cleaning Lab also works with formulators and distributors of cleaning supplies, as well as commercial and community organizations interested in promoting a healthier environment for the population. Specifically, the work of the Cleaning Lab aims to reduce the use of solvents that contribute to ozone depletion, global warming, and volatile organic compound (VOC) emissions. In addition, decreases in exposures to flammable, carcinogenic, and other toxic substances are targeted.

When choosing alternatives, the Cleaning Lab knows that the process is challenging and is careful to not shift the risk from the worker to the environment or vice versa. To help make this selection process easier, prior to performance testing, the lab conducts a preliminary screening of products for health and safety issues. After completing the testing to determine how well a product works, the Cleaning Lab follows the preliminary safety screening with a more in-depth analysis to assess any potential negative impact for worker and public health, using its assessment tool, the Pollution Prevention Options Analysis System (P2OASys).

The Cleaning Lab has performed thousands of testing trials on hundreds of products for end users, formulators and suppliers of cleaning products (including equipment). The Cleaning Lab uses standardized testing procedures for testing both industrial cleaning and janitorial cleaning products. The Lab works with a company's existing conditions (substrates, soils, contaminants, and cleaning equipment) in the lab and on location when needed. The on-site piloting with a company's equipment and staff has become the lab's signature action to make sure that the product and the process work well together. This hands-on technical assistance is what takes the theory of lab work to the reality of the end users.

Even when the lab work and piloting on-site gets completed, there are other challenges to overcome. Acquiring buy in from management is one such obstacle. Likewise, gaining the acceptance from existing clients of the new process must be achieved to ensure successful adoption of the alternatives to TCE. Acceptance may require extensive documentation and validation that the process will meet stringent specifications from military, aerospace, and FDA applications. Establishing a diversified project team will help to highlight and address other concerns related to the specific project being considered.

This book will provide a basic background on cleaning before moving onto the pros and cons of using TCE. Next, the steps to take to identify process specific alternatives and how to evaluate these products for performance and safety will be presented. A discussion follows from there on what has been evaluated at the TURI Cleaning Lab. Several additional resources for identifying and evaluating safer alternatives beyond the lab are reviewed throughout the text, as well as obstacles to overcome that have been identified by the TURI Cleaning Lab in its twenty years of experience. How these obstacles have been addressed and overcome will be laid out to provide further assistance to those seeking the TCE silver bullet.

Chapter 2

Background on Cleaning

Abstract

The groundwork behind why cleaning is important and how it has been accomplished over the past 40+ years. Solvents have been developed, tested, implemented and then abandoned due to unintended worker health and environmental safety concerns.

Every book or chapter on surface cleaning starts with a definition of what "clean" means. And this is for good reason, as cleaning can mean many things to many people. An easy way to define it would be to say clean means free from dirt, stain, or impurities. Or even more simply, unsoiled. Soils can be defined as any extraneous or unwanted material deposited and/or attached to a surface and cleaning is the process of getting rid of these impurities.

Industrial surface cleaning is performed for a number of reasons. It may be required to prepare the surfaces of parts prior to other manufacturing processes such as welding, plating or painting. It may be performed for aesthetic reasons as an aid for marketing and sales. Or it may be necessary to ensure that the finished product will perform without failure caused by contamination [6].

Another source defines surface cleaning as soil management...cleaning work is about moving chemical materials from where they are not wanted to where they are wanted [7]. A final example of the definition of cleaning is the removal of a sufficient amount of soil to allow adequate performance of the product, obtain acceptable visual appearance as required, and to achieve the desired surface properties [8].

All cleaning systems depend on three actions: mechanical, thermal, and/or chemical. For each cleaning process, it is a balancing act of these three actions, the latter having the most relevance to the focus of this book. With a good chemical cleaner, the mechanical and thermal requirements can be lowered. From the chemical vantage point, solvency is an important characteristic. Solvency can be defined as the ability to dissolve. Water is considered, by many, to be the 'universal solvent,' capable of dissolving many inorganic and some organic contaminants or soils.

But not all soils readily dissolve in water alone, which is why additives were included to make the first soaps. A standard dictionary definition has soap defined as a cleansing and emulsifying agent made usually by action of adding alkali on fat or fatty acids and consisting essentially of sodium or potassium salts of such acids. The earliest literary reference to soap dates from the third millennium BC [6].

Further enhancements in cleaning came about from the introduction of detergents, also known as synthetic soaps, which were introduced in 1930 and were found to perform better than soaps in hard (i.e., mineral-laden) water because they contained water softeners to effectively treat magnesium and calcium carbonates.

A step in the wrong direction saw the development of complex phosphates following World War II, primarily sodium tripolyphosphate. These chemicals increased the cleaning power of detergents but the 'phosphate builders' were found to contribute to the accelerated eutrophication of rivers, lakes, and streams throughout Europe and the Americas since they contained phosphorus, an essential nutrient for algae and other water plant growth.

In addition to soaps, some terpenes, natural organic compounds occurring in the essential oils and oleoresins of plants (lemon, orange) and conifers (balsam, pine), have solvating powers and therefore are used in a wide array of cleaning applications. As unsaturated hydrocarbon polymers of isoprene (C5H8), terpenes can be constituents in semi-aqueous cleaners in order to improve upon cleaning performance and or environmental safety.

The abilities of early soaps and detergents were limited. The principle of 'like dissolves like' could be demonstrated by simple visual observation in the search for cleaning products. The natural soaps and detergents simply did not dissolve greases and oils, common industrial contaminants, on their own. Therefore, further research was needed to improve cleaning.

Halocarbon chemistry played an important role in the development of more successful synthetic cleaners. Carbon tetrachloride was first produced in Germany in 1839 and marketed under the trade name Katharin as a grease remover [5]. Even though it was effective on fats and oils, asphalt, rubber, bitumens, and gums, and it was notable as a nonflammable solvent for many materials sold in solution, its vapors were found to be highly toxic. This was not the silver bullet, so the search continues for a safe and effective solution.

While most of the synthesized compounds possessed refrigerant or propellant properties of the Freon series, some, like CFC-113, exhibited characteristics suitable for cleaning needs. Attempts were made to make the finished products ‘safer,’ but the hazardous nature of their common chemical feedstock was apparent. In addition to the feedstock being unsafe, these finished products had an unfortunate environmental hazard, ozone depletion. Not the silver bullet, therefore the search continues.

Other halogenated compounds containing fluorine or bromine demonstrate good cleaning ability but represent no improvement in health or environmental considerations. Chlorofluorocarbons (CFCs) differ from chlorinated hydrocarbons by the insertion of fluorine. Hydrochlorofluorocarbons (HCFCs) are similar to CFCs except that some of the halogen atoms are substituted with hydrogen. These were marketed as less toxic replacements for current ozone depleting substances. However, these products did not eliminate the compounds’ ozone-layer damaging effects. HFCs are another potential CFC alternative. Although they contain no chlorine, their high vapor pressure and low solubility make them poor cleaners. In addition, the CFCs and HCFCs contribute to global warming [5]. Still no silver bullet, the search continues.

Carbon tetrachloride (CCl4) is an example of a simple halocarbon. Other halogenated compounds containing fluorine or bromine demonstrate good cleaning characteristics but result in no improvement in health or environmental considerations. Furthermore, carbon tetrachloride is also a volatile organic compound (VOC). Evaporating into the atmosphere at ambient conditions, VOCs include almost all cleaning solvents. Ozone, the primary component of smog, can be formed when VOCs are released and undergo photochemical reactions with atmospheric oxides of nitrogen. Not the silver bullet.

Hydrofluoroether (HFE) is a halocarbon that showed potential as a replacement solvent for methyl chloroform and CFC-113 in metal, electronic, and precision cleaning applications, but has a Global Warming Potential (GWP) of 150-480 (the higher the GWP the worse the effect is on the environment, carbon dioxide has a GWP of 1). Perfluorocarbons (PFCs), which contain only carbon and fluorine atoms, also exhibit good cleaning properties and are extremely inert. They were not viewed as a danger to the ozone layer, but their atmospheric lifetimes are thought to be thousands of years and so these compounds would most likely have a very strong potential for enhancing global warming. Holster that gun, still searching for the silver bullet.

Brominated hydrocarbons, such as dibromomethane, for example, were also used for cleaning, although on a much smaller scale. On the downside, bromination may be several times more damaging to atmospheric ozone than chlorination.

Hydrocarbons can be derived from petroleum, coal, and natural gas. They are compounds containing only carbon and hydrogen molecules. Aliphatic or fatty acid-based hydrocarbons such as mineral spirits are used in cleaning operations.

Organic solvents, the traditional solvents for the removal of organic materials, include alcohols (e.g., methanol), benzene, turpentine and ether. Another compound, methyl ethyl ketone, deserves mentioning in this class of solvents, a favorite for cleanup of paints, oils, resins and celluloses.

Trichloroethylene (TCE) is of major commercial significance in metal cleaning and is the major focus of the book. The uses and hazards are covered in more detail throughout the next few chapters.

The dry-cleaning fluid PCE is used because of its non-flammability, high solvency, vapor pressure, and stability. Dichloromethane (CH2Cl2), also known as methylene chloride, is another colorless, nonflammable liquid that is soluble in alcohol and is used in paint removal, dewaxing, and degreasing, as well as a refrigerant. Besides immersion, vapor degreasers are employed with this type of industrial cleaning, but their mechanical action can be viewed as secondary because of these solvents' excellent solvating powers. In the vapor phase cleaning process, the final rinse of the parts being cleaned is achieved by condensing solvent vapors on the parts. The next chapter will cover the basics of vapor degreasing in a little more detail.

One would think that these chlorinated solvents would be the ideal candidates for the end game of an effective product. But, these chemicals are associated with negative environmental and health effects which include atmospheric ozone depletion, global warming, acid rain formation, carcinogenicity, neurotoxicity reproductive toxicity, and cardiovascular and central nervous system damage. Blow the smoke away, not the silver bullet either, but we keep on trying.

To address these health effects, there was an attempt to return to the way things were, that is, to use the universal solvent, water, albeit with some additional ingredients. As mentioned previously, surface cleaning or degreasing can be defined as the removal of contamination, or unwanted material, from a surface. The process of lifting soil from a surface by displacing it with surface-active materials that have a greater affinity for the surface than they do for the soil is known as detergency. Differing from the solvents, water-based or aqueous cleaners may not depend on solvent penetrability for their cleaning efficiencies. Aqueous cleaners containing detergents are more complex and are usually mixtures containing surfactants, alkaline, or other 'builder' materials, and sometimes non-aqueous components. Cleaning with these products relies on a combination of processes such as: solubilization, wetting, emulsification, deflocculation, sequestration, and saponification. As one can imagine, these products typically are safer than solvents but are not as robust as the TCE and other solvents. This limitation exists despite water's universal solvent moniker. The water-based cleaner is often hampered by its surface tension, which affects its ability to remove soil from small blind holes. In addition, for some surfaces, if water is not removed properly, the excess solution may lead to formation of rusting, machining fines being left in these holes. These pitfalls can cause rejected parts or problems down the manufacturing line, and the ever present possibility that business will be lost if the client is not happy with the quality of the cleaning..

Semi-aqueous cleaning was an attempt to incorporate the principles of aqueous and organic cleaners. To make this happen, one would combine a surfactant with a low-volatility hydrocarbon such as a terpene to form a cleaning blend. Unlike traditional vapor degreasing, cleaning with semi-aqueous cleaners does not rely on boiling liquids nor is it restricted to a constant boiling composition. In its simplest form, semi-aqueous cleaning involves two steps: (1) the organic component to dissolve the soil, and (2) the water component to remove the residues of the blend and any other water-soluble soils. The surfactant ensures the water-solubility of the otherwise insoluble hydrocarbon.

Remember from the beginning of our discussion, cleaning was defined simply as removal of unwanted material from a surface. However, achieving a clean surface is not so simple. While it may be possible to identify an aqueous/semi-aqueous cleaning product that will replace the solvent TCE, there is no silver bullet that works in every situation, all of the time. To make matters even more challenging, sometimes it is not the cleaning that is the biggest issue. For some situations, the steps following cleaning can be more challenging. These actions would include rinsing, drying, and the next manufacturing step in the facility. In some cases, having a little oil or grease on a part that is going to sit for three months before being used would not be a bad thing. But if you were going to be painting, plating or welding the part, the excess soil or any remaining water would be detrimental.

Ultimately, a process is necessary to help you identify your needs, determine the level of cleanliness desired, and establish what types of products to try and test them in both the lab (theory) and in the field (practice) to confirm the alternative will work best for your specific needs. But more will be discussed on this topic in later chapters.

Chapter 3

WHY TCE

ABSTRACT

The solvent trichloroethylene continues to be used due to its cleaning versatility. It is very forgiving to the end user in terms of what it can remove and how well it can do so. Not a lot of thinking needed to go into the cleaning of manufactured parts. A review of the chemicals physical and performance characteristics is reviewed and helps to set the stage for why a universal drop-in has not been identified.

Bottom line, it works. For many it was considered the Silver Bullet for cleaning. TCE is a highly versatile solvent that works on a lot of soils and end user applications. TCE is an effective solvent for a variety of organic materials. The solvent is also forgiving of operator inexperience, easy to use, and relatively inexpensive. In addition, for military, aerospace, and other stringent processes it is a well known and proven solvent. When it was first widely produced in the 1920s, TCE's major use was to extract vegetable oils from plant materials such as soy, coconut, and palm. Other uses in the food industry included coffee decaffeination and the preparation of flavoring extracts from hops and spices. It has also been used for drying out the last bit of water for production of 100% ethanol.

Today, the most common cleaning use of TCE is to remove grease from fabricated metal parts and some textiles. Other applications the solvent is used in includes ingredients in adhesives, paint removers, typewriter correction fluids, rug-cleaning fluids, spot removers, pepper sprays and manufacturing of batteries. In1991 a report of U.S. production stated a production volume of 321 million pounds of TCE [9].

WHERE IT WORKS

In solvent cleaning applications, the TURI Cleaning Lab has worked with more than 40 companies in a wide spectrum of industry during a twenty year span. Table 3-1 lists the general categories that the companies fall into and a count of companies with which the Cleaning Lab has worked. As is evident, most of the work conducted (more than 75%) has been for metal working and general manufacturing.

Table 3-1. Company Types Using TCE

Companies	Total 47
Capacitor Manufacturer	2
Plating & Metalworking	22
Aircraft	1
Semiconductor/Electronics	2
General Mfr	14
Jewelry	4
Tools	1
Wire & Cable	1

As expected, when dealing with metal working companies, the major type of soils cleaned are oils (57%) which include lubricants and other cutting fluids. In addition, buffing compounds (30%) are another category that has been successfully cleaned using TCE. The general types of soils cleaned using TCE are shown in the Table 3-2 on TURI Lab documentation. The table also includes the number of specific contaminants identified by the companies.

Table 3-2. Contaminant Cleaned Using TCE

Contaminants	Tests Run	Specific Contaminants
Buffing/Polishing Compounds	226	12
Oil	423	33
Paints/Inks/Coatings	62	9
Waxes	15	9
Other	14	4

As illustrated from the work at the TURI Cleaning Lab, TCE has been seen by many companies as being a very effective cleaner and many would be more than satisfied to continue using the solvent. From the January 2013 EPA draft risk assessment on TCE, specific industry sectors were identified. Table 3-3 lists the sectors listed using the North American Industrial Classification System (NAICS) codes. The list is not intended to be exhaustive, but rather provides a sampling of the potentially affected entities linked to the usage of TCE.

Table 3-3. Industry Classifications Potentially Using TCE [10]

NAICS code	Industry Sector
314	Textile Product Mills
321	Wood Product Manufacturing
323	Printing and Related Support Activities
325	Chemical Manufacturing
326	Plastics and Rubber Product Manufacturing
331	Primary Metal Manufacturing
332	Fabricated Metal Product Manufacturing
333	Machinery Manufacturing
334	Computer and Electronic Product Manufacturing
335	Electrical Equipment, Appliance, and Component Manufacturing
336	Transportation Equipment Manufacturing
337	Furniture and Product Related Manufacturing
339	Miscellaneous Manufacturing
488	Clothing and Clothing Accessory Stores
493	Warehousing and Storage
811	Repair and Maintenance
928	National Security and International Affairs

A majority of these companies are using vapor degreasing applications. In this process, parts are cleaned in a continuous unsoiled solvent that can get into hard to reach areas. Vapor degreasing uses the solvent to dissolve the contaminants on the part and remove them by dripping off the part. The solvent condenses on the colder work-piece and then the liquid solvent dissolves the greases on the part. With the impurities contained in the liquid beads, the solvent runs off the part. A typical vapor degreasing unit with spray wand and ultrasonic transducers is depicted in Figure 3-1.

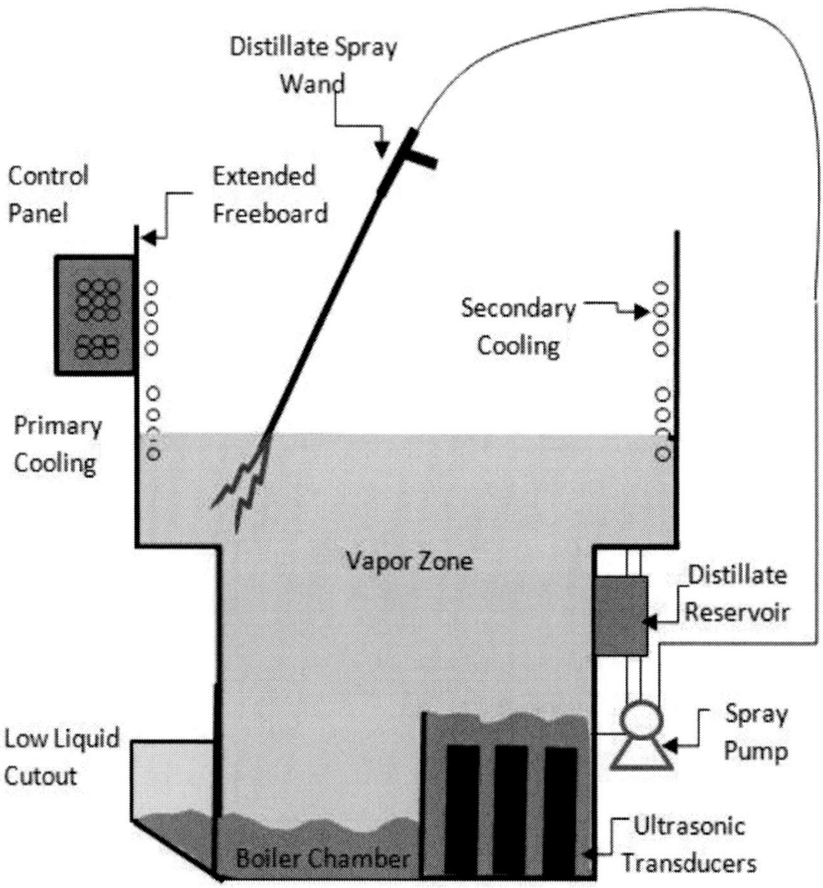

Figure 3-1. Vapor Degreasing Unit with Spray and Ultrasonic.

How It Works

With its physical characteristics, TCE is highly suited for this type of cleaning process. Many of the solvents characteristics are listed in Table 3-4 below.

Table 3-4. Physical Properties of TCE

Physical Property	Vapor Pressure	Boiling Point	Latent Heat	Surface Tension	Flash Point	SNAP Approved	KB Value	Hansen Solubility Parameters		
Units	mm Hg	C	cal/g	dynes/ cm	C			δd	δ p	δ h
Perchloro-ethylene	15.8	121	50.1	32.32	none	Yes	90	18.3	5.7	0
Methylene Chloride	355	40	78.9	28.2	none	Yes	136	17.0	7.3	7.1
TCE	69	87.2	57	29	none	Yes	125	18.0	3.1	5.3
Water	23.8	100	539	72.8	none	NA	6.5	15.5	16.0	42.3

Vapor pressure is an indication of a liquid's evaporation rate, the higher the vapor pressure the more readily it will be in the vapor phase. From the table, perchloroethylene and TCE have lower vapor pressure than methylene chloride, resulting in easier evaporation from the surface being cleaned for methylene chloride.

Tied to vapor pressure is the boiling point. The normal boiling point is the temperature at which the vapor pressure equals the ambient atmospheric pressure. With any incremental increase in that temperature, the vapor pressure becomes sufficient to overcome atmospheric pressure and lift the liquid to form vapor bubbles inside the bulk of the substance. TCE has a lower boiling point than perchloroethylene and water and would require less energy to reach operating conditions.

Latent heat is the amount of energy absorbed or released by a substance during a change in its physical state that occurs without changing its temperature. With a lower latent heat, a solvent would require less energy to convert from a liquid to a gas. Water requires a significant amount of energy to convert from its liquid phase to the gas phase, an order of magnitude higher than TCE.

Surface tension is a contractive tendency of the surface of a liquid that allows it to resist an external force. Cleaning agents that have lower surface tension make it a better "wetting agent" to get into and out of pores, blind holes, and fissures rather than bridging them with surface tension. So if cleaning were first to take place in a bath of TCE, its lower surface tension would help to clean out much of the soil from dirty parts. Many times, solvent cleaning will utilize an immersion cleaning followed by the vapor degreasing process all within a single enclosed cleaning system.

The flash point of a volatile material is the lowest temperature at which it can vaporize to form an ignitable mixture in air. Like water, TCE and the other listed solvents in the table have no flash point which would reduce the risk of fire that some of the other organic solvents have.

The Significant New Alternatives Policy (SNAP) Program is EPA's program to evaluate and regulate substitutes for the ozone-depleting chemicals that are being phased out under the stratospheric ozone protection provisions of the Clean Air Act (CAA) [11]. One of the driving forces of the TURI Cleaning Lab has been to identify solvents that do not cause damage to the ozone layer.

Kauri-butanol value (Kb) is an international, standardized measure of solvent power for a hydrocarbon solvent. Mild cleaners have low scores in the tens and twenties; powerful cleaners like the listed chlorinated solvents have ratings in the low hundreds. The Kb process is a test and is well-suited to traditional industrial cleaners. In the past, using a Kb value to compare industrial cleaners like methylene chloride, perchloroethylene, and TCE made sense because of the types of parts which were being cleaned (big) and the types of contamination they were trying to remove (crude) [12]. In the place of the Kb test, most engineers are substituting systematic testing procedures, rigorously testing solvents on their components, their contamination, and their cleaning processes. This makes sense because it is a test of real-world conditions.

Another way to assess a solvent's cleaning potential without implementing direct testing is to use the Hansen Solubility Parameters (HSP). This process is a way of predicting if one material will dissolve in another and form a solution. The parameters are based on the idea that like dissolves like where one molecule is defined as being 'like' another if it bonds to itself in a similar way. Specifically, each molecule is given three Hansen parameters, each generally measured in $MPa^{0.5}$. The first looks at the energy from dispersion forces between molecules, δd. Second is the energy from dipolar intermolecular force between molecules, δp. The third is related to the energy from hydrogen bonds between molecules, δh.

The goal would be to match a solvent's three parameters as closely as possible to the solubility parameters of the soil to be removed. Typically, there is not an exact match, and compromises must be made. Through the use of equations designed to assist in identifying the best fit, the end user can be successful in finding solvents or a combination of solvents that should be suitable for the application. It does not replace the need for process specific testing on actual parts under typical operating conditions, but it is a place to start to help narrow down the options for a particular process, part, and soil.

As mentioned, both Kb and HSP will not be perfect indicators of what will work for a cleaning situation. The importance of testing alternatives on process specific soils and parts in the lab and piloting in the field is significant. However, both can be used to narrow the list down to a more manageable level more quickly than a trial and error method.

For TCE, its high stability in the presence of common chemical stabilizers and low boiling point permit vapor degreasing with low heat input. These characteristics add to how easily it can be recycled. The solvent combines a broad cleaning capacity with greater economy than many other solvents. For some, it is the "devil" who they know and with whom they are "comfortable." It has been used for decades and was the solvent accepted for many military, aerospace, and high-end applications. These are many of the reasons why it is still being used despite the many environmental, health, and safety reasons to not use it.

Chapter 4

WHY NOT TCE

ABSTRACT

Trichloroethylene despite its effectiveness in cleaning has been determined to have significant health and environmental risks associated with it. It is one of the most common chemicals identified at hazardous waste sites across the United States. In addition to the environmental concerns, the chemical has both acute and chronic health effects.

As demonstrated, TCE is a solvent that has broad effectiveness for degreasing operations. For that reason, the chemical has wide spread use across the country. With the increase in usage, so increases the potential for exposure to the chemical. When a substance is released from a large area, such as an industrial plant, or from a container, such as a drum or bottle, it enters the environment.

Wide usage for many years has led to widespread contamination from TCE. The Environmental Protection Agency (EPA) has identified 1,428 hazardous waste sites as the most serious in the nation. These sites make up the National Priorities List (NPL) and are targeted for long-term federal clean-up. Trichloroethylene has been found in at least 861 NPL sites or more than 60% of these locations. This may be a conservative estimate of the breadth of TCE in the environment as it is unknown how many NPL sites have been evaluated for TCE. As EPA looks at more sites, the sites with trichloroethylene may increase [13].

A snap shot of some of the relevant regulatory information on TCE is listed in Table 4-1.

Table 4-1. TCE Regulatory Information [14]

U.S. FEDERAL REGULATORY INFORMATION
EPA has set a Maximum Contaminant Level (MCL) for drinking water of 0.005 mg/L. Regulates drinking water for public water systems and drinking water wells that serve at least 25 people.
U.S. ATSDR has set Minimal Risk Levels (MRL) for inhalation and oral routes.
One of 188 hazardous air pollutants (HAPs) listed under section 112(b) of the 1990 Clean Air Act Amendments and is regulated for more than 170 industrial source categories.
Ranked as number 16 out of 275 substances on the 2005 Priority List of Hazardous Substances for the Comprehensive Environmental Response, Compensation, and Liability Act (CERCLA) section 104(i), as amended by the Superfund Amendments and Reauthorization Act (SARA).
Emergency Planning and Community Right-to-Know Act (EPCRA) Section 313 "Toxic Chemicals," quantities of TCE greater than 25,000 pounds manufactured or processed, or greater than 10,000 pounds otherwise used, must be reported; under the Comprehensive Environmental Response, Compensation, and Liability Act (CERCLA), reporting releases of TCE of any quantity exceeding 100 pounds is required.

It is no surprise that one of the biggest sources of trichloroethylene in the environment is evaporation from factories that use it to remove grease from metals. Other avenues for releases into the air and water takes place when the chemical is disposed of at chemical waste sites. Even though it evaporates easily in air, TCE can stay in the soil and in groundwater for extended periods of time. Once it is in the air, about half will be broken down within a week, which seems like it would be a good thing. If it breaks down, then it is no longer present, no more worries, right? Not so fast. The drawback is that when TCE brakes down, it forms more harmful chemicals in the air such as phosgene, a lung irritant.

In 2011, global consumption of TCE was 945 million pounds with expected growth at about 1.5% annually over the next five years. The corresponding U.S. consumption was 255 million pounds [15]. TCE production volume of 224.7 million pounds was reported to EPA in the 2012 Chemical Data Report (CDR).

According to research from EPA on the use of TCE, over 4.2 million pounds of TCE are released into the air from both large and small sources. See

Table 4-2 on annual air emission data. Of the large sources, the vast majority of the emissions are from open top vapor degreasers. The EPA estimates that approximately 30,000 workers and occupational bystanders at small commercial degreasing operations are exposed to TCE emissions.

Table 4-2. TCE Emission Sources and Corresponding Total Annual Air Emissions of TCE as reported in the 2008 NEI

Type of Emission Source	Number Sources	Total Annual Air Emission (lbs/yr)
Point Source (large facility)	180 (degreasing units)	1,480,000
Nonpoint Source (small facility)	1779*	2,860,000

*Nonpoint sources are aggregated and reported at the county level. Thus, the number of nonpoint sources (as reported in NEI) will not necessary correspond to the number of degreasing units.

Breakdown of Degreasing Machine Type based on NEI Data for Point Source

Type of Degreasing Machine	Number Units	Total Annual Air Emissions (lbs/yr)
Open-top Vapor Degreasing (batch vapor)	116	890,000
Cold Solvent Cleaning (cold)	13	140,000
Conveyorized Vapor Degreasing (in-line)	11	120,000
General Degreasing Units (unknown)	40	330,000

In the body, TCE may break down into dichloroacetic acid (DCA), trichloroacetic acid (TCA), chloral hydrate, and 2-chloroacetaldehyde. These products have been shown to be toxic to animals and are probably toxic to humans.

Exposure to TCE has been reported to have adverse effects on the central nervous, immune, and endocrine (hormonal) systems in adults. Reported effects often involve the central nervous system, with reported symptoms of fatigue, sleepiness, headache, confusion, and blurred vision. Other effects on liver, kidney, gastrointestinal tract, and skin have been reported in addition to enhancing the negative effects of alcohol in humans, allowing less intake of alcohol to cause larger impairment in the person drinking the alcohol. Available data suggest that exposure to TCE is associated with cancer of the kidney and other organs. The U.S. EPA Integrated Risk Information System (IRIS) reassessment of TCE is ongoing and includes analysis of noncancer and cancer data [14].

In support of the IRIS, the U.S. EPA completed a toxicological review of trichloroethylene in September 2011. The resulting document was the culmination of efforts spanning more than twenty years. TCE is carcinogenic to humans by all routes of exposure and poses a potential human health hazard for non-cancer toxicity to the central nervous system, kidney, liver, immune system, male reproductive system, and the developing embryo/fetus [16].

Human Health Effects Acute Exposure

Acute (short-term) exposure to TCE vapors can cause central nervous system (CNS) effects (e.g., light-headedness, drowsiness, headache, and giddiness) and may lead to unconsciousness or be fatal. At very high exposure levels, TCE can sensitize the heart to the effects of adrenaline and similar agents, which may lead to sudden cardiac arrest. In addition, TCE may irritate the respiratory tract at high vapor concentrations, and may be lethal at concentrations exceeding 1000 ppm. Repeated or prolonged contact with the chemical in liquid or mist form can cause irritation of the skin and eyes. Direct exposure to liquid TCE defats the skin, causing redness, blistering, and scaling.

Chronic Exposure

Frequent, prolonged exposures to organic solvents in general can have long-lasting and possibly permanent effects on the nervous system. Chronic exposure to TCE in excess of recommended occupational limits has been associated with damage to the liver and kidneys, although this is less well documented in humans than in animals. People who drink water containing TCE over many years could experience liver problems and may have an increased risk of developing liver cancer. TCE also has genotoxic and immunotoxic potential, and some studies indicate that it may be a teratogen.

REPRODUCTIVE/DEVELOPMENTAL EFFECTS

Evidence from animal and epidemiologic studies suggests that several reproductive and developmental toxicity end points may be associated with TCE exposure, including infertility in males and females, impaired fetal growth, and cardiac teratogenesis. The U.S. Agency for Toxic Substance and Disease Registry Public Health Statement for TCE, however, does not conclude that there is a direct relationship between exposure to TCE and an increased risk of reproductive effects.

ENVIRONMENTAL HAZARDS

TCE most often enters the environment via fugitive emissions from metal degreasing industries and by spills or accidental releases to air, soil, and water. Because of its high vapor pressure and low rate of adsorption in soil, TCE evaporates fairly rapidly when released to soil. However, where it persists it can percolate rapidly through sandy soil thereby reaching groundwater. Chemical and biological degradation of TCE in water are expected to be very slow. TCE is not expected to accumulate in aquatic organisms or to adsorb

Table 4-3. Summary of TCE Health Effects

ACUTE (SHORT TERM)	CHRONIC (LONG TERM)	CANCER HAZARDS
• Inhaled, acts as CNS depressive, and exposure can cause dizziness, unconsciousness, visual disturbances, nausea, and vomiting. • Exposure can result in skin irritation and corneal injury. • GI absorption is rapid if ingested and can produce CNS depression. Aspiration onto lungs can cause chemical pneumonia. • Very high levels can cause irregular heartbeat which can be fatal.	• Repeated contact can cause irritation, blistering, roughening, and cracking of exposed skin. • Repeated exposure can cause memory loss, headache, depression, and weakness in the arms and legs. • Drinking or ingesting large amounts may cause liver and kidney damage and impaired immune function. • Breathing for long periods of time may cause impaired heart function, unconsciousness and death.	• It may cause mutations and possible impaired fetal development. • There is evidence that TCE causes lung kidney and liver cancer in animals. • It may cause lung and liver cancer. • SOURCE: Product MSDS's and Tech Data Sheets

onto sediment; however it is toxic to aquatic organisms. When released to the atmosphere, TCE remains in the vapor phase. Once in the air about half of the TCE will break down within a week. When TCE is broken down in the air, phosgene, a significant lung irritant, can be formed. In the body, TCE may break down into dichloroacetic acid, trichloroacetic acid, chloral hydrate, and 2-chloroacetaldehyde. These byproducts have been shown to be toxic to animals and may be toxic to humans [17].

Table 4-3 lists a summary of the acute and chronic effects as well as the cancer risks of TCE.

SUMMARY

TCE has been listed by the EPA as a probable human carcinogen and a possible reproductive hazard (teratogen). Acute exposure to TCE affects the central nervous system, the kidneys, and the liver. Because of its potential health effects, this chemical is regulated throughout its lifecycle. Many uses are heavily restricted due to its toxicity. In 2009, the Commonwealth of Massachusetts designated TCE to be a Higher Hazard Substance, and requires users to report use of over 1000 pounds per year and to plan for a safer alternative.

There are many situations where alternatives can be found for TCE. The challenges are about finding the willingness to spend capital, move to something safer or overcoming the needs of their clients that can hold up the process.

Chapter 5

IF NOT TCE, THEN WHAT

ABSTRACT

In Massachusetts, the use of TCE has declined in large part due to the efforts of the state's toxics use reduction program. The research lab specifically established to assist in cleaning related substitution projects has been working with companies for more than two decades. During that time, the methods developed have been proven to be very effective in assisting companies through the solvent replacement process.

There are those that think they cannot do what they do without their reliable solvent. Solvents have been used for decades in all fields of cleaning. Many of these solvents, even though they work well, are regulated due to the environmental and worker health issues. Efforts have been made to reduce worker and environmental exposure. More often than not, these changes arise due to the variations in legal structures, especially liability issues, and account for how most societies approach environmental decision-making for cleaning applications [18].

Furthermore, many pollution prevention assistance providers consider this field to be "conquered" and have moved on to other niches to focus their efforts on the next battle. In many cases, particularly among larger companies, uses and emissions of TCE have been minimized (but not eliminated, so the work continues) through substitution and other engineering controls [19]. To illustrate, from the first year of reporting for the Toxic Use Reduction Act (1990) through 2005, companies in Massachusetts have reduced their use of TCE by more than 77%. Most companies were able to reduce their use of TCE to below the 10,000 pound reporting threshold within five years of the beginning of the program. The trend in reduction of TCE in the Commonwealth continues when looking at the reporting amounts in 2010. During this year, the data show the overall reduction to be in excess of 90%.

Inputs include TCE that is manufactured or processed, as well as TCE that is "otherwise used" (i.e., uses that do not become incorporated into the final product). Outputs include TCE that is generated as byproduct (i.e., all non-product material created by a process line prior to release, on-site treatment, or transfer) and the amount of TCE that is shipped in or as product. Both inputs and outputs of TCE were significantly reduced overall in the Commonwealth from 1990 to 2010. Table 5-1 lists the TCE usage in Massachusetts.

Table 5-1. Reported TCE Data in Massachusetts

Year	Total Used	Byproduct	Shipped	Releases
1990	3,651,227	2,389,120	1,129,237	1,309,690
2005	834,462	74,726	587,001	31,388
2010	294,836	103,278	209,754	50,555
Year	Processed	Otherwise Used	Total Used	
1990	1,215,555	2,435,672	3,651,227	
2005	596,416	238,046	834,462	
2010	203,052	91,784	294,836	

Specifically, the total input of TCE, including otherwise used, manufactured, or processed TCE, was reduced by 91.9% and the total output of TCE, the amount generated as byproduct, shipped in or as product, over the same time period also was reduced by a similar amount, 91.1%. Figure 5-1 shows the reduction of TCE usage in Massachusetts during the 20 year period.

This makes a strong case for having conquered the TCE in cleaning conundrum. However, even with the dramatic reduction in reported TCE usage, the number of companies filing on this usage has decreased by 61% (41 to 16). In addition, of the original 41 companies reporting for TCE usage in the TURA program, only four still reported in 2010. It is important to note, that not reporting does not equate to not using. These companies have reduced their usage to below the reporting level, or have reduced their work force below the 10 employee level required for reporting.

Between 1990 and 2012 there have been 99 companies in Massachusetts that have reported using TCE. For those companies that reported on TCE usage for 10 or more years, the highest percentage of applications has been degreasing at 36%, followed closely by the repackaging/ blending sector, 32%.

If the otherwise used category gets combined with degreasing (as many of the companies were determined to be using TCE for cleaning based on interactions of the TURI Cleaning Lab with these companies), the degreasing/cleaning usage surpasses 54%. The remaining sector of TCE usage includes adhesives and paint manufacturing, accounting for about 13%. Table 5-2 breaks down the usage categories for the top 22 companies reporting on TCE for more than 10 years in Massachusetts.

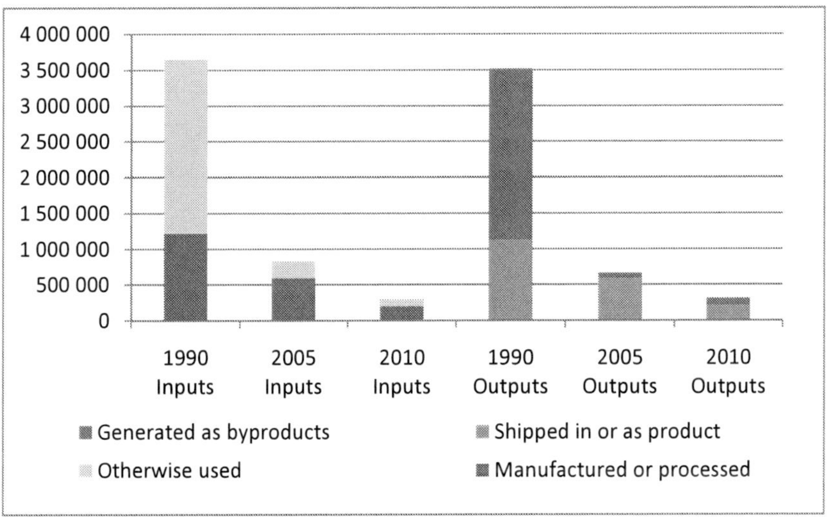

Figure 5-1. TCE Reduction in Massachusetts.

Table 5-2. Massachusetts Companies Reporting More than 10 Years on TCE Usage

Usage	Subgroup	# Companies	% usage
Repacking/Blending		7	31.82
Paints/Adhesives		3	13.64
Degreasing		8	36.36
	Degreasing	5	
	Manual	1	
	Vapor	2	
Otherwise Used		4	18.18
Combined Cleaning Usage			54.55

Continuing the case that all is not conquered in pursuit of replacing TCE for cleaning, of the 16 companies reporting on TCE usage in 2010, 62% (10) reported using TCE for cleaning. The other two groups each had 18% (3) of the companies.

Table 5-3. Example Standardized Cleaning Protocols

American Society for Testing and Materials (ASTM)	B322-85 Standard Practice for Cleaning Metals Prior to Electroplating A 380 Cleaning, Descaling, and Passivation of Stainless Steel C 756-87 Standard Test Method for Cleanability of Surface Finishes D 1133 - Kauri-Butanol Value of Hydrocarbon Solvents D 4261-83 Standard Practice for Surface Cleaning Concrete Unit Masonry for Coating D 6361-98 Standard Guide for Selecting Cleaning Agents and Processes D 5703-95 Standard Practice for Preparatory Surface Cleaning for Clay Brick Masonry E 1548-93 Standard Practice for Preparation of Aerospace Contamination Control Plans E 1575-98 Standard Practice for Pressure Water Cleaning and Cutting E 1857-97 Standard Guide for Selection of Cleaning Techniques for Masonry, Concrete, and Stucco Surfaces F 502-93 el Standard Test Method for Effects of Cleaning and Chemical Maintenance Materials on Painted Aircraft Surfaces G 121 Preparation of Contaminated Test Coupons G 122 Evaluating the Effectiveness of Cleaning Agents G 127-95 Standard Guide for the Selection of Cleaning Agents for Oxygen Systems G 131-96 Standard Practice for Cleaning of Materials and Components by Ultrasonic Techniques
Military/ Aerospace	MIL PRF 85570 Performance Specification: Cleaning Compounds, Aircraft, Exterior MIL PRF 87937 Performance Specification: Cleaning Compound, Aerospace Equipment MIL-PRF-680 Performance Specification: Degreasing Solvent MIL-PRF-29602A Performance Specification: Cleaning Compounds, Parts Washer And Spray Cabinet ARP 1755 Effect of Cleaning Agents on Aircraft Engine Materials Stock Loss Test Method
3rd Party Certifiers and Other Standards	CCD-146 Hard surface Cleaners DFE standard-for-safer-cleaning-products Ecolabel CCD-110 Cleaning and De-greasing Compounds GS-34 - Green Seal Standard for Cleaning And Degreasing Agents GS-53 Standard Specialty Cleaning Products for Industrial and Institutional Use

Identifying a substitute for solvents in cleaning applications is not an easy task. There are thousands of formulations from which to choose. The variability of literature for these products changes from vendor to vendor as well as from product to product, making the search for an applicable substitute nearly impossible. Even after selecting a potential product, there is no guarantee that it will work with the company's specific soils, parts and process.

So how do you make sure it works? The problem with the way most people evaluate cleaning product effectiveness is that their comparisons tend to be subjective, not scientific. However, well-written test protocols and repeatable science need to play a key role in choosing cleaning products that work consistently well. A partial listing of cleaning standards from various organizing bodies is listed in Table 5-3.

Part of the TURI Cleaning Lab's mission is to test and evaluate the effectiveness of different cleaning chemistries and equipment on a variety of substrates and soils. The many years of experience of the TURI Cleaning Lab has led to being able to evaluate products using methods that have been tailored to the specific end user's needs. Many times the standardized methods listed in the table are structured for a limited application. Being able to understand a process that falls outside the standardized process is critical. The essentials of the standard can then be adjusted to provide a more realistic comparison of possible alternative products. Another adaptation of the standard is to include on-site testing to verify results obtained in the lab with field testing.

Prior to beginning a project, a brief questionnaire should be filled out so that testing can be tailored to meet the individual firm's needs. The survey can also assist plant personnel involved with cleaning to become familiar with all aspects of the process. The two main areas on which to gather information include the part to be cleaned and the process used to clean. Table 5-4 lists some of the relevant details to collect.

The parts to be cleaned should be described by the material of which they are made, the complexity, size, and shape. When possible, having sample parts before and after cleaning will help to provide bench mark levels of cleanliness. For the process being used, identifying the soil to remove is the most important element. Without this, it will be next to impossible to select an alternative cleaner. Collecting samples of the soils will allow for process specific testing to take place. The current type of cleaning is important but in the end, the process might change. How long cleaning takes place, including rinsing and drying times, helps to define the acceptable window for operating the cleaning process. One element that should not be overlooked is who does the cleaning. Identifying the staff involved and including them in the assessment process will go a long way to making a successful switch.

Table 5-4. Questions to ask about cleaning processes

Part to be cleaned	
What is it made of?	Aluminum, brass, ceramic, copper, electronics, glass, marble, nickel, stainless steel, steel, plastic, titanium, zinc
Surface type	Smooth, rough, hard, or soft
Geometry	Simple, flat, complex containing inaccessible areas
Size	Small, medium, large; dimensions
Weight	< 1/2lb, < 1lb, < 5lb, 10lb, < 50lb, > 50lb weight: Minimum/Maximum
Samples available	Yes/No. Before and after cleaned
Process Being Used	
What needs to be removed?	Oil, lubricant, cutting fluid, grease, wax, buffing compound, adhesive, resin, paint, ink, dirt, grime
Samples available	Yes/No
Amount of soil used	Per part, Monthly, Annually
Amount cleaned	Per week, per batch
Process sequence	What manufacturing step is done immediately before cleaning? After cleaning?
Equipment used	Vapor degreaser, air sparging, immersion, dip tank, sink on drum, spray cleaning, ultrasonic, manual wiping, manual scraping, other mechanical agitation
Chemical/solvent to be replaced	Manufacturer of chemical, volume used in equipment, amount used per year. How long does cleaning last. What is the operating temperature?
Rinsing/Drying	Does the part need to be rinsed, dried? What are the current steps?
How clean is clean?	How is the cleaning process assessed? Visual, wipe, gravimetric, contact angle, painting success rate
Who does cleaning?	How many people are involved, how much training do they have/receive?

In addition to how and what are being cleaned, it is important to learn the purpose of the cleaning (i.e., desired product specifications) and what (if any) problems are present with the current cleaning system. Many times working with the company's supply chain is necessary. In these cases, cleaning may be restricted in the specifications for the parts being manufactured. This means that after an alternative solvent or solvents is found, specific client based testing may be required for the change to be approved. This has often been the case with military and aerospace projects. Likewise, medical devices or human implants that are overseen by the FDA encounter similar obstacles.

Having collected as much information on the process as possible, the standardized cleaning protocols reviewed, and the customized testing methods identified, the next step is to identify what the possible alternatives could be.

Chapter 6

IF NO SILVER BULLET, THEN HOW

ABSTRACT

Tools were developed to assist in the search for safer alternatives. Many of them were designed to compare health and safety features of the products, while others provide performance information. The chapter lists various attributes of each tool and compares and contrasts with the other tools. Working from this information, The TURI Lab constructed a more comprehensive database that would look at EHS and performance and be able to provide a list of possible products for specific situation.

Typically, replacement of trichloroethylene (TCE) has been on a case-by-case basis and an all encompassing systematic, scientific approach has been generally lacking. Over the past twenty years several tools were developed to assist in the search for safer alternatives. In 2007 a comparison of several tools was conducted by the TURI Cleaning Lab. The tools are listed in Table 6-1. Sadly, of the dozen that were reviewed, only five still remain available in 2015. Of these, two offer the user specific products that could be used for cleaning. Two others provide the end user with a way to assess health and safety of a product but do not provide a list of what to compare.

As part of the review of these tools in 2007, a new database was developed by the TURI Cleaning Lab and has since expanded to include more and more testing results.

Table 6-1. Solvent Substitution Tools Evaluated in 2007

Solvent Substitution Related Tools	Web Address in 2007	Available 2015
SAGE – Solvent Alternative Guide	http://clean.rti.org	No
Joint Services Pollution Prevention Technical Library	http://p2library.nfesc.navy.mil/	No
ISSDS - Integrated Solvent Substitution Data System	http://es.epa.gov/issds/	No
Joint Services Solvent Substitution Database	https://js3.ctc.com/	No
GreenBiz Toolbox	www.greenbiz.com/toolbox/	No
Paris II - Program for Assisting the Replacement of Industrial Solvents	www.tds-tds.com/parfact.htm	Yes
CleanTool	www.cleantool.org/	Yes
Indiana Chemical Hazard Scoring System	http://cobweb.ecn.purdue.edu/CMTI/IRCHS/	No
P2OASys – Pollution Prevention Options Analysis System	http://www.turi.org/p2oasys	Yes
Aqueous Way to Go	www.angelfire.com/band2/ greencleaners/doctoralthesis.html	No
GreenScreen	http://www.greenscreenchemicals.org/method	Yes
System developed in 2007		
CleanerSolutions	www.cleanersolutions.org	Yes

PROGRAM FOR ASSISTING THE REPLACEMENT OF INDUSTRIAL SOLVENTS (PARIS II)

The software tool was set up to allow the user to design less hazardous solvents without compromising solvent performance. It brought together a set of unique algorithms developed under the leadership of Dr. Heriberto Cabezas of the National Risk Management Research Laboratory, USEPA, located at Cincinnati, Ohio [20]. The advanced technology and extensive knowledge base of PARIS II makes it easy to identify "green" replacements for solvents currently in use in far less time than previously possible. The system includes chemical data and estimation methods, thus the user needs to know only the composition of the solvent to be replaced in order to run the program.

The PARIS II tool was intended to make designing new solvents for single chemicals or mixtures easier. In addition, evaluations of solvents for their impact on the air and the environment can be performed. A review of toxicity, health, and safety effects allows the user to fully review and compare possible alternative solvents. It makes it possible to match new solvents to existing chemical processes. The system provides a ranked list of solvents to expedite the replacement of dangerous solvents.

The PARIS II system was designed to aid the user in the identification of possible replacement materials for current solvents. The tool also offers the ability to design mixtures when no single compound can meet user specifications. Unfortunately, a lot of physical characteristics of solvents are needed in order to obtain useful alternative information. Further, the complete tool must be purchased in order to get any useful information on the possible alternatives.

CleanTool

From a European Innovation Project, the CleanTool database was developed with a goal of creating a database of best practice for the cleaning of metal surfaces. Objective information, collected from industries across the European Union was included in the tool, allowing enterprises to select the cleaning process that best fits their business. The database was released at the end of 2004.

Metal cleaning can often be seen as a basic task that requires little thought or consideration, yet substantial benefits may be gained by taking a fresh look at this process. According to the project managers, the CleanTool database should allow users to make the best choice of cleaning application by providing comprehensive information on key areas [21].

Throughout its existence, new technologies will provide information on many cleaning processes and agents. A performance section covers measurement tools for assessing the effectiveness of a cleaning process. Cost effectiveness information has been provided on issues such as the cost of technology or the operating life of a cleaning agent. An environmental impact section of the database gives information on the effect a process has on the environment. A section on health and safety will present information to allow the risk to workers to be assessed and reduced by means such as the substitution of a substance with one less hazardous.

CleanTool has the potential for being a good source of process specific information, providing specific answers to questions surrounding metal cleaning processes. The database does address many topics other than cleaning. However, at the time of the review, the tool only looked at metal working industries and had a limited collection of information.

The tool provides a summary of processes covering costs, technology, quality, occupational safety and health, and environment. The results can be sorted by each category. Each process is ranked as very good, good, satisfactory, sufficient, bad, and no statement for the five areas. Each is color coded for easy visual assessment.

The tool does include process information and gives alternative name and manufacturer contact. However, the user has to go to the process sheet and click on tabs to find product names/vendor contacts. To do so, several layers of the website must be gone through before getting to any specific information on products.

POLLUTION PREVENTION OPTIONS ASSESSMENT SYSTEM (P2OASYS)

This program was designed to be a systematic tool that helps companies determine whether the toxics use reduction (TUR) options being considered may have any unforeseen negative environmental, worker, or public health impacts. Specific process information would be utilized to establish a baseline level of hazards associated with the current system (solvent). The chemical information of potential replacements would then be inputted and compared directly with the original product. P2OASys provides the end user with a numerical hazard score for the inputted current process chemical as well as the identified options. The resulting information would allow the user to combine this data with other information sources and professional expertise to make decisions on TUR option implementation.

The tool will assist companies in two ways. The first allowed for a systematic examination of the potential environmental and worker impacts of TUR options in a comprehensive manner, examining the total impacts of process changes, rather than simply those of chemical changes. The second way compared TUR options with the user's current process based on quantitative and qualitative factors. Cleaning alternatives to be evaluated must be first identified through another source, as this tool does not contain any specific products.

P2OASys tool provides the user with a lot of information on safety for both workers and the environment. However, it was not designed to propose an alternative cleaning product, as the user must provide the alternatives compared in this tool. The program can take a while to use as well because considerable effort is needed to fill out the matrix with alternative product information. However, once data has been entered, the tool does provide for a consistent way to compare a current solvent with possible alternatives. One of the benefits of this tool is the ability to use both quantitative and qualitative data on the chemical toxicity, ecological effects, and physical properties of both current and proposed chemistries. Table 6-2 lists the categories and criteria used in the P2OASys tool.

Table 6-2. P2OASys Criteria

Category	Key Criteria
Acute Human Effects	Inhalation Median lethal concentration (LC50)
	Permissible Exposure Level (PEL)/ Threshold Limit Value (TLV)
	PEL/TLV (dusts/particles)
	Immediately Dangerous to Life or Health (IDLH)
	Respiratory Irritation
	Oral Median Lethal Dose (LD50)
	Dermal Irritation
	Skin Absorption
	Dermal LD50
	Ocular Irritation
Chronic Human Effects	Reference Dose (RfD)
	Carcinogen
	Mutagen
	Reproductive Effects
	Neurotoxicity
	Developmental Effects
	Respiratory- Sensitivity/Disease
	Other Chronic Organ Effects
Physical Hazards	Heat
	Noise Generation
	Vibration
	Ergonomic Hazard
	Psychosocial Hazard

Table 6-2. (Continued)

Category	Key Criteria
Aquatic Hazards	Water Quality Criteria (HWQC)
	Aquatic LC50
	Fish No Observed Adverse Effect Concentration (NOAEC)
	Plant Half Maximal Effective Concentration (EC 50)
	Observed Ecological Effects
Persistence/ Bioaccumulation	Persistence
	Biochemical Oxygen Demand (BOD) Half-Life
	Hydrolysis Half-Life
	Bioconcentration (Octanol-Water Partition Coefficient)
	Bioconcentration Factor (BCF)
Atmospheric Hazard	Greenhouse Gas
	Ozone Depletory (ODP)
	Acid Rain Formation
	National Emission Standards for Hazardous Air Pollutants (NESHAP)
Disposal Hazard	Landfill
	Emergency Planning and Community Right-to-Know Act (EPCRA) Reportable Quantity
	Incineration
	Recycling
Chemical Hazard	Vapor Pressure
	Solubility in Water
	Specific Gravity
	Flammability
	Flash Point
	Reactivity
	pH
	Corrosivity
	High Pressure System
	High Temperature System
	Mixture/Reaction Potential
	Odor Threshold
	Volatile Organic Compound (VOC)
Energy and Resource Use	Non Renewable Resource
	Water Use
	Energy Use
Product Hazard	Upstream Effects
	Consumer Hazard
	Disposal Hazard
Exposure Potential	Exposure Potential

Environmental Health and Safety (EHS) Methodology

The Green Screen for Safer Chemicals defines a path to chemicals that are safer for humans and the environment. It is a rigorous, hazard-based screening method that is designed to inform decision making by businesses, governments, and individuals concerned with the risks posed by chemicals and to advance the development of green chemistry [22].

The elements of the Green Screen process are similar to those in the P2OASys tool. Of the 19 total elements in the Green Screen, 15 appeared in the P2OASys. Benchmarks are unique to Green Screen and are a major strength of the method. Each benchmark is determined by analyzing specific combinations of hazard classifications. The benchmarks were developed to reflect hazard concerns that have been established by governments nationally and internationally. Each hazard classification is used in applying the benchmark criteria to determine the final benchmark score. It is possible that a benchmark cannot be assigned at all if the data are insufficient [23].

Benchmark 1—Avoid—chemicals of high concern
Benchmark 2—Use but search for safer substitutes
Benchmark 3—Use but still opportunity for improvement
Benchmark 4—Prefer—safer chemicals

The criteria for each benchmark become progressively more demanding for environmental and human health and safety, with the hazard criteria of Benchmark 4 representing the most preferred chemicals. The hazard level of these criteria will determine into which benchmark the chemical falls. The more hazardous a chemical, the lower the benchmark in which the chemical will belong.

With the provided categories, the first step in the process would be to determine chemicals to evaluate. With the chemicals identified, research was needed to collect data and begin assessing chemicals. This was accomplished by examining comprehensive toxicological data, checking Green Screen specified lists, and using estimated data from suitable analogs or modeled data where measured data are lacking for the parent chemical. With the data collected, next in the process would be to classify the hazard level for each hazard endpoint (e.g., vH, H, M, L, vL).

Table 6-3. Green Screen Criteria

Grouping	Hazard Endpoints
Human Health Group I	Carcinogenicity (C) Mutagenicity and Genotoxicity (M) Reproductive Toxicity (R) Developmental Toxicity, including Neurodevelopmental Toxicity (D) Endocrine Activity (E)
Human Health Group II	Acute Toxicity (AT) Systemic Toxicity & Organ Effects (ST-single) Neurotoxicity (N-single) Skin Irritation (IrS) Eye Irritation (IrE)
Human Health Group II*	Systemic Toxicity and Organ Effects, Repeated Exposure Sub-Endpoint (ST-Repeat) Neurotoxicity – Repeated Exposure Sub-Endpoint (N-Repeated) Skin Sensitization (SnS) Respiratory Sensitization (SnR)
Environmental Toxicity and Fate	Acute Aquatic Toxicity (AA) Chronic Aquatic Toxicity (CA) Persistence (P) Bioaccumulation (B)
Physical Hazards	Reactivity (Rx) Flammability (F)

Like P2OASys, this method was not designed to propose an alternative cleaning product, as the user must provide the alternatives compared in this tool. The process can take a while to use and takes considerable effort to fill out the matrix with alternative product information.

CleanerSolutions – A Bullet of Almost Silver Status

In 1993, TURI established the Surface Cleaning Laboratory in order to address the need to provide assistance in finding safer cleaning alternatives. The TURI Cleaning Lab was designed with the capability to evaluate the effectiveness of different cleaning chemistries and equipment for a variety of substrates and contaminants. The goal was set to assist industry in the search for safer cleaning processes by developing and promoting safer alternatives to hazardous solvents with a special focus on aqueous/semi-aqueous cleaners and state-of-the-art surface cleanliness analyses.

From the testing performed at the TURI Cleaning Lab, a wealth of data has been accumulated on the performances of industrial cleaning products. In order to use this information effectively, a database system was created so that this data could be accessed easily and quickly. The database was created after extensive review of other tools, surveys of potential end users, and trainings on possible tools were conducted. The end goal was to allow users to search for an alternative cleaning product based on safety and effectiveness.

Table 6-4. CleanerSolutions Safety Screening

VOC Content Range (g/l)	GWP	ODP	NFPA/HMIS Assessment	pH range
0-24	GWP = 0	ODP = 0	Examples	0-1.0
25-49	GWP = 1 (CO_2)	All others > 0	H-0 F-0 R-0	1.1-2.4
50-74	All others		H-0 F-0 R-1, H-0 F-1 R-0	2.5-2.9
75-99			H-1 F-1 R-0, H-2 F-0 R-0	3.0-4.0
100-149			H-1 F-1 R-1, H-2 F-1 R-0	4.1-5.9
150-199			H-3 F-0 R-0	6.0-6.4
200-299			H-2 F-2 R-0, H-1 F-2 R-1	6.5-7.5
300			H-1 F-3 R-0	7.6-8.9
> 300			H-2 F-2 R-1	9.0-9.9
			H-1 F-3 R-1, H-2 F-3 R-0	10-11.4
			H-2 F-2 R-2	11.5-11.9
			H-3 F-3 R-0	12-12.4
			H-3 F-3 R-1, H-3 F-3 R-2	12.5-12.9
				13-14

The most important consideration when choosing an alternative is to not shift the risk involved in the process. For example, when attempting to replace a flammable solvent, it would not be advantageous to select an ozone depleting chemical. In this case, the risk goes from the worker (burn, explosion, etc.) to the environment (ozone destruction). Ideally, a selection of an alternative cleaning product should be safer for one or the other, or if possible, a product that would result in a reduced risk for both. Therefore, in keeping with the objective of the TURI Cleaning Lab to develop and promote environmentally-friendlier, safer alternatives to hazardous solvents, the determination of a chemical's overall risk is based on the following environmental indicators: volatile organic content (VOC), global warming potential (GWP), ozone depletion potential (ODP), flammability/reactivity and toxicity/carcinogenicity. Specifically, the environmental scoring is based on the values for VOC, GWP, ODP, National Fire Protection Association (NFPA) or Hazardous Materials Identification System (HMIS), and pH.

With a screening method to quickly assess EHS criteria, the next step is to evaluate how well the products work. To do this the CleanerSolutions database includes the number of trials performed for specific contaminants (both effective and ineffective). Specific search information will be entered into a database query, generating a list of products matching inputted criteria. The number of effective and ineffective trials associated with each listed product is likewise displayed along with the safety screening score, thus providing safety and performance with one tool, using one search.

How Do They Compare?

A survey was generated to ask potential end users what makes an online tool useful when looking for an alternative cleaning product. The surveys were administered to environmental and occupational experts. The questions were rated for each database on a scale of 1 (worst) to 5 (best). The total score was the summation of all 21 questions. A maximum score of 105 was possible (21 questions multiplied by the highest score of 5). In some instances, the tools evaluated could not be assessed which resulted in a zero being entered for that question. The maximum score for these products were compared to the number of questions answered [24]. Of the tools evaluated in 2007, no single tool was shown to meet all of the needs of an online tool for identifying an alternative cleaning product. From these assessments, the CleanerSolutions database was created to help fill the gap of a universal tool for solvent substitution for cleaning. When evaluating CleanerSolutions using the same questions, the tool was a clear favorite, scoring an 88% favorable rating. CleanTool was found to have a rating of 74%. Table 6-5 contains results of the surveys for the four remaining online tools for solvent substitution (non-online tools that are still available were not included).

Table 6-5. Survey of Tools

Questions to answer	P2OASys	Paris II	CleanTool	CleanerSolutions
Ease of Use	3	2	2	5
Easy to Understand Results	4	3	4	4
Meaningful Results	4	3	3	5
Specific Lists	0	0	3	5
Directed Searches	0	0	4	4
Quick Results	0	0	3	5
Detailed Searches	0	0	4	3
Simple Searches	0	0	3	5
Product Information	0	0	3	5
Vendor Contact	0	0	3	5
Proven Performance	0	0	4	4
Health and Safety Information	5	5	4	4
How Products were Tested	0	0	5	5
How Products Compare w/Current Product-Process	5	5	4	5
Demonstrations	1	3	4	4
Instructions – Descriptions	3	3	4	4
Contact Information	3	5	5	5
Help Request	3	5	5	5
Glossary	2	3	5	5
Save Searches – Comparisons	4	3	4	3
Printability	5	3	4	2
Total	42	43	78	92
Max Score	60	60	105	105
% Rating	70	72	74	88

While the tool was being developed, there were several presentations, workshops, and hands-on trainings. Each of these provided useful insight into how the CleanerSolutions database tool was viewed by potential end users. The creation of the tool was greatly enhanced by these sessions, ensuring that the tool would meet the needs of industry personnel, EHS officials, technical assistance providers, and regulatory agencies.

The current working site contains several search mechanisms, from the very simple to moderately detailed. Most use of the database has been on the simple side. The most basic search method requires the inclusion of the contaminant to be removed. The search method can have the surface material (substrate) and process (equipment) added to further refine a search.

To illustrate, if the database was being used to search for a product to clean lubricants, cutting fluids, and/or grease, the current query would yield 5,592 records. Nearly 6,000 records is a lot to sift through to make an informed selection. Adding a substrate, say aluminum, to further define the search, cuts the list down to 1,322 records. By adding in possible equipment, ultrasonics, the search can be culled down more to just 68 records. A final filter for effective products cuts the list down to a manageable 48 records. Figure 6-1 shows the search field filled out for all of the added variables. This iterative search process allows the user to get a sense for the vast amounts of possible alternatives that exist even in this one tool from one resource. The method only takes a minute or two to produce a workable number of records.

Once the list is generated (see Figure 6-2 Search Results) more information can be obtained by drilling deeper into the database. Clicking on a product name will reveal data about the product as supplied by the vendor and any laboratory generated results for the product (performance and health and safety) as shown in Figure 6-3.

In addition to the basic search for alternatives, the CleanerSolutions site was designed to provide other avenues for looking for alternative solvents. Using the same simple click and select methodology, the *Replace a Solvent* search allows the end user to investigate options based on tested alternative chemistry for specific solvent cleaners. The user can choose a solvent from the menu bar and add in the contaminant, surface materials, and cleaning method to provide a listing of possible products that have previously been tested for replacing the selected solvent.

In some instances, the search can be controlled not by direct substitution for a solvent based on performance, but by the need for safety. Throughout the existence of the database tool, many requests have come in asking for a list of safer alternatives based on the various health and safety criteria. With that in mind, the CleanerSolutions site added the *Safety Screening Search* method, shown in Figure 6-4. The screening score is based on 50 points, with a 10 point maximum for each parameter listed. A higher safety screening score implies a potentially safer product. In the search interface the user can filter by specific end points for the five EHS data. Or the database can be used to look for a range of final safety screening scores instead. Many times the solvents to be replaced fall below a safety score of 37 and the search method allows the user to filter out all the products below that point.

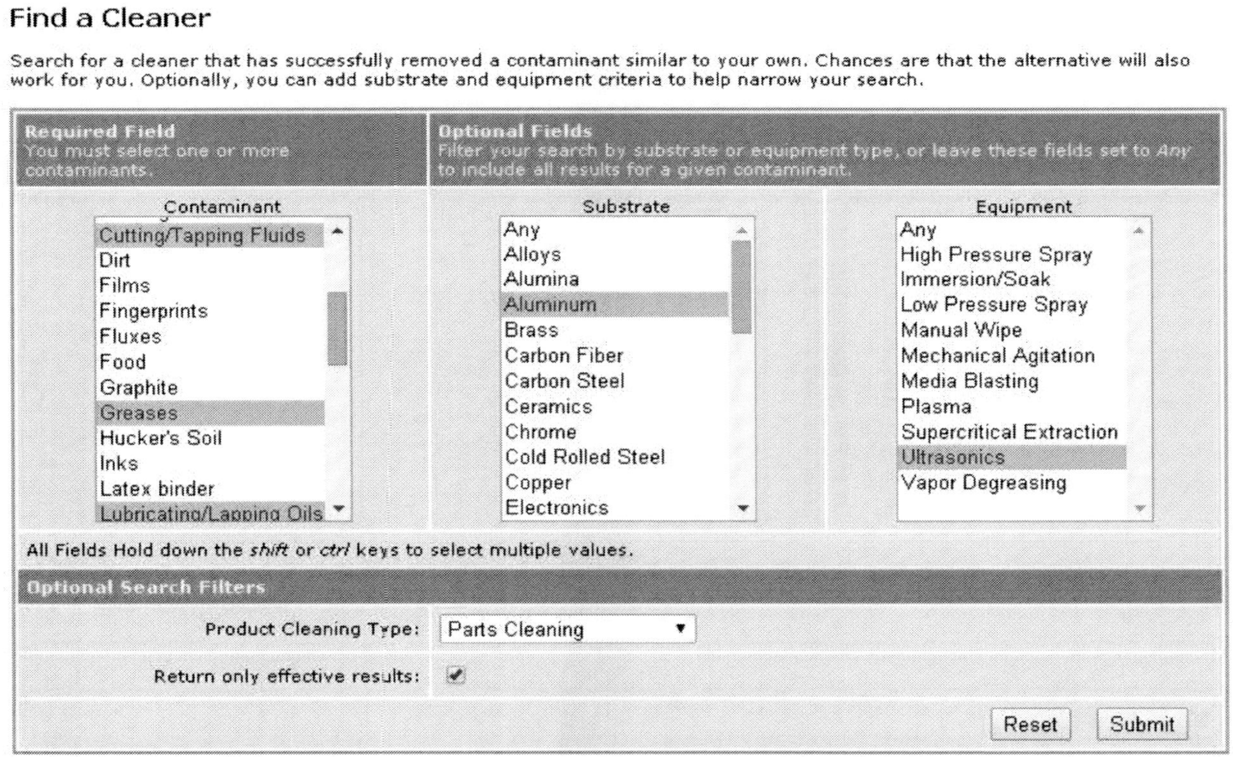

Figure 6-1. Search Method Refined.

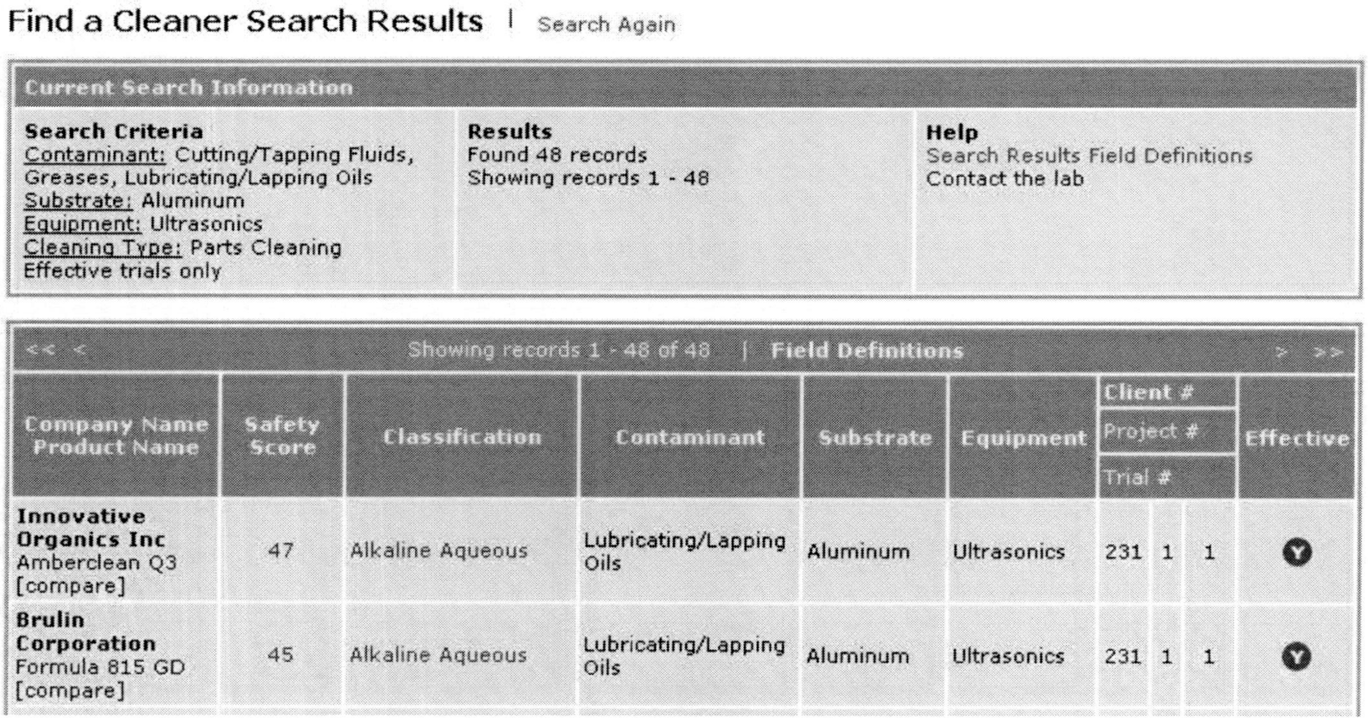

Find a Cleaner Search Results | Search Again

Current Search Information

Search Criteria	Results	Help
Contaminant: Cutting/Tapping Fluids, Greases, Lubricating/Lapping Oils Substrate: Aluminum Equipment: Ultrasonics Cleaning Type: Parts Cleaning Effective trials only	Found 48 records Showing records 1 - 48	Search Results Field Definitions Contact the lab

<< < Showing records 1 - 48 of 48 | Field Definitions > >>

Company Name Product Name	Safety Score	Classification	Contaminant	Substrate	Equipment	Client #	Project #	Trial #	Effective
Innovative Organics Inc Amberclean Q3 [compare]	47	Alkaline Aqueous	Lubricating/Lapping Oils	Aluminum	Ultrasonics	231	1	1	Y
Brulin Corporation Formula 815 GD [compare]	45	Alkaline Aqueous	Lubricating/Lapping Oils	Aluminum	Ultrasonics	231	1	1	Y

Figure 6-2. Search Results.

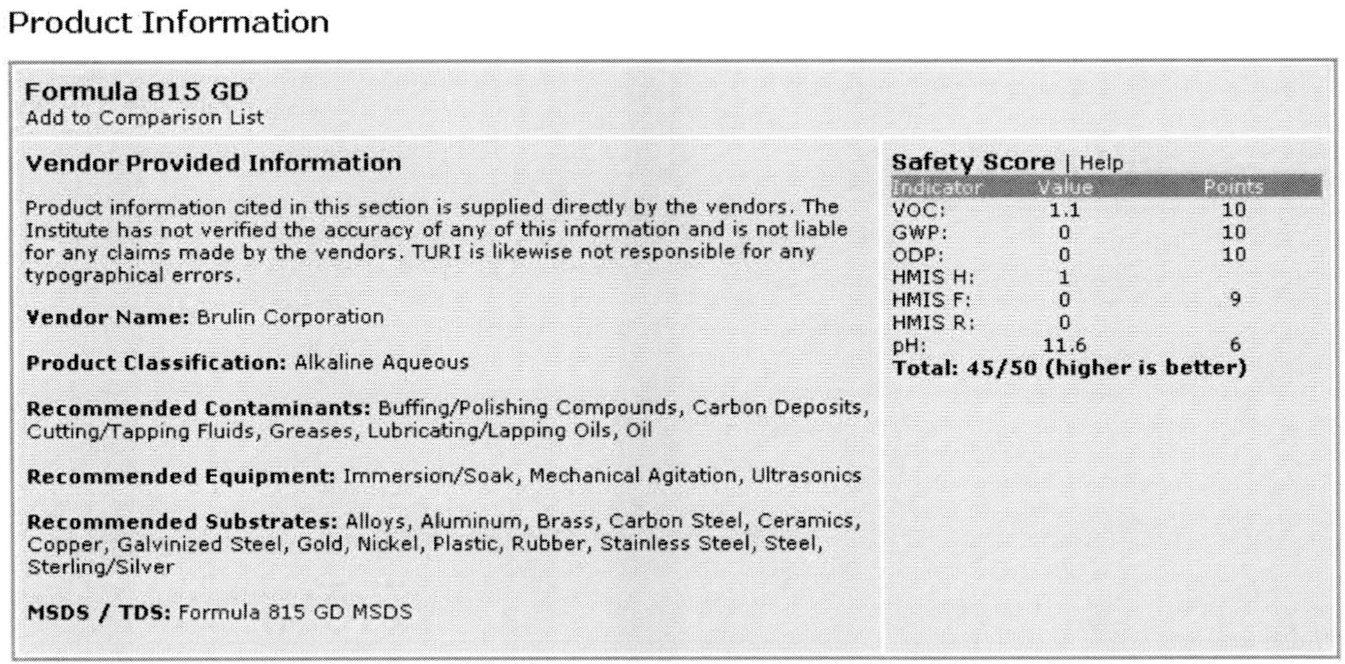

Product Information

Formula 815 GD
Add to Comparison List

Vendor Provided Information

Product information cited in this section is supplied directly by the vendors. The Institute has not verified the accuracy of any of this information and is not liable for any claims made by the vendors. TURI is likewise not responsible for any typographical errors.

Vendor Name: Brulin Corporation

Product Classification: Alkaline Aqueous

Recommended Contaminants: Buffing/Polishing Compounds, Carbon Deposits, Cutting/Tapping Fluids, Greases, Lubricating/Lapping Oils, Oil

Recommended Equipment: Immersion/Soak, Mechanical Agitation, Ultrasonics

Recommended Substrates: Alloys, Aluminum, Brass, Carbon Steel, Ceramics, Copper, Galvinized Steel, Gold, Nickel, Plastic, Rubber, Stainless Steel, Steel, Sterling/Silver

MSDS / TDS: Formula 815 GD MSDS

Safety Score | Help

Indicator	Value	Points
VOC:	1.1	10
GWP:	0	10
ODP:	0	10
HMIS H:	1	
HMIS F:	0	9
HMIS R:	0	
pH:	11.6	6

Total: 45/50 (higher is better)

Figure 6-3. (Continued).

Laboratory Evaluation of Formula 815 GD | Field Definitions

Client #	Project #	Trial #	Contaminant	Substrate	Equipment	Effective
8	1	1	Stickies	Textile	Immersion/Soak	Y
13	1	0	Lubricating/Lapping Oils	Aluminum	Immersion/Soak	Y
13	1	0	Lubricating/Lapping Oils	Stainless Steel	Immersion/Soak	Y
13	1	1	None	Alloys	Immersion/Soak	Y
13	1	1	None	Aluminum	Immersion/Soak	Y
13	1	1	None	Steel	Immersion/Soak	Y
28	1	0	Oil	Aluminum	Immersion/Soak	Y
28	2	0	Oil	Aluminum	Immersion/Soak	N
28	2	1	Oil	Aluminum	Immersion/Soak	Y
28	2	2	Cutting/Tapping Fluids	Aluminum	Immersion/Soak	N

Figure 6-3. Product Information Page.

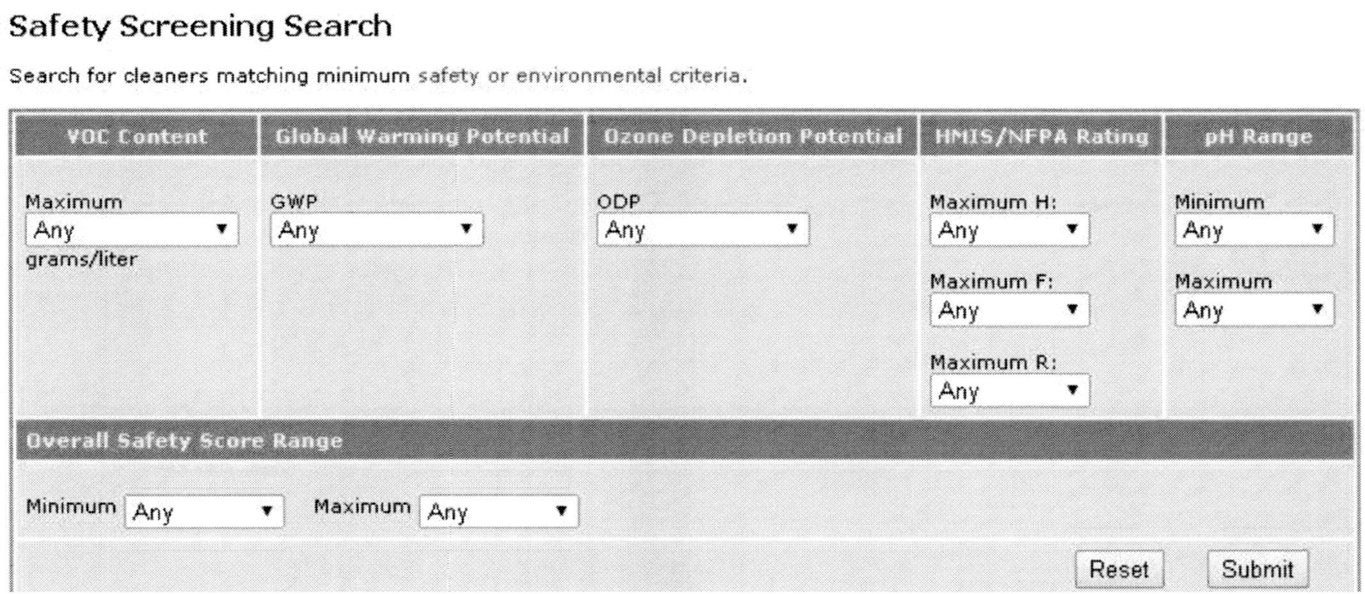

Figure 6-4. Safety Screening Interface.

Amberclean L 12 [x]	SC Soyester [x]	Smart Solve 605 [x]
Vendor Name: Innovative Organics Inc	**Vendor Name:** Gemtek Products	**Vendor Name:** United Laboratories International
Classification: Alkaline Aqueous	**Classification:** Ester	**Classification:** Biobased
Recommended Contaminants: Cutting/Tapping Fluids, Dirt, Greases, Lubricating/Lapping Oils, Oil, Waxes	**Recommended Contaminants:** Adhesive, Coatings, Cutting/Tapping Fluids, Fluxes, Greases, Inks, Lubricating/Lapping Oils, Paints, Resins/Rosins	**Recommended Contaminants:** Adhesive, Carbon Deposits, Coatings, Cutting/Tapping Fluids, Fluxes, Greases, Inks, Lubricating/Lapping Oils, Oil, Paints, Resins/Rosins, Waxes
Recommended Equipment: Immersion/Soak, Low Pressure Spray, Manual Wipe, Ultrasonics	**Recommended Equipment:** Cold Solvent, Immersion/Soak, Manual Wipe	**Recommended Equipment:** Cold Solvent, Immersion/Soak, Manual Wipe, Mechanical Agitation
Recommended Substrates: Aluminum, Brass, Carbon Steel, Ceramics, Copper, Electronics, Galvinized Steel, Glass/Quartz, Gold, Iron, Nickel, Plastic, Stainless Steel, Steel, Titanium	**Recommended Substrates:** Aluminum, Stainless Steel, Steel	**Recommended Substrates:** Alloys, Aluminum, Brass, Copper, Galvinized Steel, Nickel, Stainless Steel, Steel
MSDS / TDS: Amberclean L 12 MSDS, Amberclean L 12 TDS	**MSDS / TDS:** SC SoyEster MSDS, SC SoyEster TDS	**MSDS / TDS:** Smart Solve 605 MSDS, Smart Solve 605 TDS

Safety Screening Information

Amberclean L 12 [x] — **Safety Score** | Help

Indicator	Value	Points
VOC:	0	10
GWP:	0	10
ODP:	0	10
HMIS H:	1	
HMIS F:	0	9
HMIS R:	0	
pH:	8.4	9

Total: 48

SC Soyester [x] — **Safety Score** | Help

Indicator	Value	Points
VOC:	64	8
GWP:	0	10
ODP:	0	10
HMIS H:	1	
HMIS F:	1	8
HMIS R:	0	
pH:	NA	10

Total: 46

Smart Solve 605 [x] — **Safety Score** | Help

Indicator	Value	Points
VOC:	0	10
GWP:	0	10
ODP:	0	10
NFPA H:	1	
NFPA F:	1	8
NFPA R:	0	
pH:	NA	10

Total: 48

Figure 6-5. (Continued).

Lab Evaluation Summary		
Amberclean L 12 [x]	SC Soyester [x]	Smart Solve 605 [x]
Number of Trials: 150 112 effective/38 ineffective	**Number of Trials:** 16 10 effective/6 ineffective	**Number of Trials:** 57 33 effective/24 ineffective
Tested Contaminants: Waxes, Lubricating/Lapping Oils, Metal fines, Greases, Inks, Oil, Buffing/Polishing Compounds, Coatings, Cutting/Tapping Fluids, Abrasive, Mold Releases, Resins/Rosins, Salts, Food	**Tested Contaminants:** Resins/Rosins, Inks, Oil, Greases, Cutting/Tapping Fluids, Lubricating/Lapping Oils	**Tested Contaminants:** Coatings, Cutting/Tapping Fluids, Oil, Inks, Adhesive, Lubricating/Lapping Oils, Resins/Rosins, Carbon Deposits, Greases
Tested Substrates: Glass/Quartz, Aluminum, Stainless Steel, Steel, Brass, Copper, Nickel, Ceramics	**Tested Substrates:** Liquid, Galvinized Steel, Steel, Aluminum	**Tested Substrates:** Steel, Aluminum, Galvinized Steel, Glass/Quartz, Ceramics
Tested Equipment: Immersion/Soak, Ultrasonics	**Tested Equipment:** Immersion/Soak, Manual Wipe	**Tested Equipment:** Immersion/Soak, Ultrasonics, Manual Wipe

Figure 6-5. Product Comparisons.

Client Information

Client Number 231 (Manufacturers of Precision Parts and Assemblies)

Client Images:

Project Number 1

Summary: Several trials have been performed for aqueous and non-aqueous based cleaning products. Supplied parts have been cleaned using both methods and sent back to the client for further analysis.

Test Objective: To reduce VOC's

Problems with Current Method: Need to reduce VOC amounts

Purpose of Cleaning: Visual specifications are critical with our components

Product Use: Analytical & chromatography components

Cleaning Chemicals: LPS Micro X, Alconox, Amberclean Q3

Trial Number	Date Run	Purpose	Success Rating
0	11/08/04	To evaluate aqueous cleaners for potential replacement of contact cleaner.	Results successful using TACT (time, agitation, concentration, and temperature, as well as rinsing and drying) and/or other cleaning chemistries examined.

Figure 6-6. Project Information.

Trial Report

Trial Number 0 (Client Number 231, Project Number 1)

Trial Purpose: To evaluate aqueous cleaners for potential replacement of contact cleaner.

Date Run: 11/08/04

Experiment Procedure:
Five products were selected from the lab's database of test results based on the products' past success in removing similar lubricants. Four of the products were diluted to 2% along with the client's current product using DI water in 600 ml beakers. A sixth product was used at full strength as recommended by the manufacturer. All six products were heated to 115 on a hot plate. Eighteen preweighed 6061 aluminum coupons were coated with Fuch's Lubricants Renocut 6515 NC (mineral oil, vegetable oil) using a hand held swab. Coupons were allowed to sit for over an hour. A second set of weights were recorded to determine the amount of soil added. Three coupons were immersed into each cleaning product and cleaned using minimal agitation provided by a stir bar. Coupons were cleaned for 5 minutes, followed by a 15 second water rinse at 120 F and airblow off for 30 seconds at room temperature. Once dry, final weights were recorded and efficiencies were calculated for each product.

Trial Results
All of the products out performed one of the client's current aqueous cleaners. Three of the products removed over 87% of the lubricant. The table lists the amount of soil added, the amount remaining and the efficiency for each coupon cleaned.

Cleaner	Initial wt	Final wt	% Removed
Amberclean Q3	0.0918	0.0481	47.60
	0.1934	0.0638	67.01
	0.1074	0.0203	81.10
Formula 815 GD	0.2651	0.0983	62.92
	0.1748	0.0528	69.79
	0.2564	0.0552	78.47
Inproclean 3800	0.1845	0.0421	77.18
	0.2610	0.0275	89.46
	0.2194	0.0459	79.08
Surface Cleanse 930	0.2272	0.0076	96.65
	0.2441	0.0255	89.55
	0.2865	0.0085	97.03
Vatron SP 2275	0.2827	0.0518	81.68
	0.2171	0.0205	90.56
	0.5620	0.0471	91.62
Ionox HC2	0.3371	0.0702	79.18
	0.3519	0.0108	96.93
	0.3171	0.0095	97.00

Success Rating
Results successful using TACT (time, agitation, concentration, and temperature, as well as rinsing and drying) and/or other cleaning chemistries examined.

Conclusion
The same six products will be retested under similar conditions. Ultrasonic energy will be added to improve efficiency.

Figure 6-7. Trial Details.

Another way the site can be utilized to assist a company is by providing a summary page of the various projects that have been worked on at the TURI Cleaning Lab. An end user could use the Browse Clients and Trials page to quickly identify other projects from a specific industry type, filtering out many of the unrelated testing results.

Any of the search results pages can provide the end user with several ways to uncover more details about the testing, safety, or product. Embedded in the results page are links to other pages on the site that expand upon the details of the executed search. Any of the columns can be arranged in either ascending or descending order (alphabetically or numerically) by clicking on the column heading. As mentioned, when clicking on some of the highlighted text a pop-up window opens providing more data or definitions of the linked item. For example, when clicking on the safety score value, the pop-up window provides the values for the five safety screening criteria for the specific product. Likewise, clicking on the classification name generates a window that gives you a definition of the various cleaning product types.

In addition to being able to open up more details on a specific product from the list as demonstrated in Figure 6-5, the tool provides the end user with a way to compare up to three products side–by-side. This feature provides a simple and quick way to choose between products as the window summarizes vendor supplied usage, safety screening scores, and performance.

In addition, by selecting a company, project, or trial number, more data is provided to the end user that will help to illustrate how the testing was conducted which is shown in Figure 6-6 and 6-7. This feature provides a deeper connection between the possible alternative and current processes with which the end user is working. The Client Information page, Figure 6-6, provides a photograph of the parts to be cleaned, a description of the overall project, why the test was being run, the problem and current method being used are, and what the product used was. All this data will help the end user assess the applicability of a project to their conditions.

Figure 6-7 is an illustration of how the testing was conducted in the laboratory setting, giving the end user a behind the scenes look of the testing methods, the other products that were included for comparison, and the next steps in the project.

SUMMARY

Without a silver bullet to use for cleaning applications, the use of the above mentioned tools can provide an end user of TCE with a way to get started. With a list of possible alternatives finally identified, the next step will be confirming that the selected products will work for the specific project. The TURI Cleaning Lab typically runs the products through a series of five steps. At each phase, the number of alternatives should be reduced to ultimately identify one or two best matches.

Chapter 7

IF NO SILVER BULLET, THEN HOW TO TEST

ABSTRACT

Because there is no silver bullet for TCE substitution, process specific testing becomes a very important part of the alternative assessment process. To ensure a product will work, a systematic evaluation process needs to be utilized. The TURI Lab has established a five step process that does just that. From product selection, through company piloting on location, the methodology has been proven to increase the likelihood of selecting an effective and safer alternative for cleaning applications.

BACKGROUND

In the TURI Cleaning Lab, the substitution of a new cleaning product for TCE is performed in a five step process. These steps include product selection, chemistry evaluation, mechanical agitation, lab piloting, and field testing. During each phase, the laboratory testing is tailored to the specifics of the current project. There is no one size fits all approach, and the TURI Cleaning Lab specializes in meeting the needs of the companies with which they work. In each phase, the number of viable products is reduced until only one or two remain, making selection by the requesting company easier and more likely adoptable at the end of the process.

Phase I Product Selection

Traditionally, this step includes monitoring the effect of drops of selected detergent concentrates on grossly contaminated coupons over time. Typically this test is needed when the soil being provided cannot be applied to a surface either because no virgin soil is available or the soil has already cured/dried. Many times this procedure can be conducted on a large number of products at a time, ranging from 10 to 40 products. Conducting a full scale evaluation on such a large number of possible alternatives can be extremely challenging. So in these mass production testing scenarios, the procedure focuses on full strength and room temperature. In some of these scenarios, a volume of the alternative product is added to a small vial and a piece of the soil (typically cured paint, resin, or some kind of solid) is then added. Observations are made at time intervals to determine what effect, if any, the solution had on the soils. Figure 7-1 shows a resin soil being subjected to the phase 1 for about twenty products.

I. Brainstorm Compatibility and "Lift" Studies

Figure 7-1 Phase I Cleaning Process.

In another option, the solution is applied directly to the soil (usually still attached to a supplied part) and allowed to sit for a short period of time. Wiping and other observations can be made to weed out the ineffective products. However, the TURI Cleaning Lab has found that more often than not, this step can be skipped due to the development of the CleanerSolutions database described in the previous chapter. The resulting list of alternatives generated from the database helps companies to select products that will most likely work for their specific situation, thus eliminating the need for the time intensive lift studies described in Phase 1. The products chosen via CleanerSolutions have been found to work in nearly 60% of the scenarios. Without the tool, products selected for testing based on reviewing product literature, past testing reports, and from the memory of lab personnel only worked in 30% of the cases [25].

PHASE II TEMPERATURE AND CONCENTRATION STUDIES

This phase and the next phase use the cleaning parameters of temperature, agitation, chemical action, and time, collectively known as TACT. If you decrease one of the four, you must compensate and increase another to keep the overall cleaning efficiency as high as possible.

How does each of these affect cleaning performance? For temperature cleaning, efficiency increases as the temperature increases. Heat makes molecules move faster, and therefore clean more quickly. Generally hotter is better; however, temperature should be adjusted according to the cleaning situation. Next, agitation, commonly known as scrubbing, is achieved through a brush and physical action, water pressure, or mechanically assisted brushes. It helps distribute cleaning agents for maximum soil suspension and removal. Third, chemical action is the process whereby the cleaning solution dissolves, emulsifies, or loosens the soil particles. Related to the chemical action is the concentration of the chemical being used. Some products may need lower concentrations to achieve the same level of cleaning. Finally, perhaps the most important aspect of the four is time. It measures the seconds, minutes, or hours for the other three TACT elements to interact for optimum efficiency. For nearly every project, time tends to be the controlling factor when discussing cleaning performance. If you cannot get it clean in the same time frame the client is currently using, in many cases it will be an uphill battle to make the case to change cleaning techniques.

To demonstrate how the balance works, here are some simplified examples. To use a cold cleaning solution would require an increase in agitation, chemical action, and/or time (the temperature variable has been decreased). To use a cleaning solution without any agitation would require an increase in cleaning solution temperature, a stronger cleaning chemical (higher concentration), and/or more time (the agitation element has been decreased). To use a water only cleaning solution would require an increase in cleaning solution temperature, more agitation, and/or more time to remove soil (the chemical action has been decreased). And to minimize time to perform a cleaning task would require an increase in cleaning solution temperature, more agitation, and/or a stronger cleaning product (the time has been decreased).

At the TURI Cleaning Lab, the first round of scientific evaluation focuses on identifying the possible effects from the chemical action alone. To facilitate the evaluation, gravimetric analysis is conducted by measuring the weight of the laboratory applied contaminant before and after cleaning to then have an empirical number representing the percent of soil removed by each potential alternative cleaner. The use of coupons, as shown in Figure 7-2, in place of actual parts allows for a more standardized, repeatable process meaning only the parameters or variables of the chemical action are in play at this time and not part geometry. Coupons typically are flat, smooth, and of the same size and shape.

Figure 7-2. Phase II Cleaning.

Coupons are selected to most closely match the surface materials of parts to be cleaned. When a company has multiple surface materials, the Cleaning Lab selects the types that are most susceptible to damage. Through the use of soils that have been identified, and in many cases supplied by the end user company and not just some prescribed soil from a test standard, testing conditions are more closely simulated to the real world scenario. During this phase of testing, time, agitation, and temperature are kept the same for all the products so that the chemical alone can become the driving force of the evaluation. As mentioned, during this phase, the number of products being compared is kept to a workable number, usually between six and ten products. Testing with coupons is done in triplicate, resulting in working with 18 to 30 coupons at a time. At the end of the testing, the more effective products will be retained for the next set of experiments addressing other elements of the TACT sequence.

PHASE III MECHANICAL ENERGY STUDIES

With a successful identification of a handful of chemical products from Phase II, the next step at the TURI Cleaning Lab is to enhance cleaning of the products by introducing a form of mechanical cleaning (agitation). The selection of equipment will depend on what the company has available or is interested in switching to. This can include manual scrubbing, ultrasonic energy (See Figure 7-3), or spray washing. Again, other TACT elements are kept constant for the products.

III. Mechanical Energy Studies
Number of chemical cleaner candidates further decreases from Phase II

Figure 7-3. Phase III Cleaning.

As in the second phase, laboratory soiled coupons are again used to assess performance. Depending on the success of the chemical assessment, time, temperature, and concentration of the process can begin to be optimized during this phase. Ideally, the desired performance should be achieved at minimum conditions. This will help to avoid increasing the cost of operation for the facility. Being able to clean at low temperature, low concentration all while achieving a short timeframe works to offset the cost of any mechanical agitation. Upon the completion of this phase of the evaluation process, the number of possible alternatives should be narrowed down to three to four products. Operating conditions of the TACT variables at this time should be well established.

PHASE IV ACTUAL PRODUCT CLEANING STUDIES

The next cleaning phase is concerned with the significance of part shape (screw configurations, inaccessible areas, line of sight cleaning, and blind holes) on cleaning. Note that in this model, mechanical energies (Phase III) are critically examined only after the proper cleaners, concentrations, and temperatures (Phase II) are achieved. Other testing protocols exist that may be as effective, but it is essential that the experimental design be logical and consistent with the end user's requirements. Assessment during this phase of the project is the first time that real parts are utilized (introducing the reality of testing). In most cases, the coupon testing of the earlier phases tend to only be an approximation of what the real world parts are like with respect to metal composition, soil loading, and aging.

So why is the early testing not conducted with real parts? It usually comes down to the fact that the number of parts that would be needed during preliminary testing far exceeds the availability. The company would need to supply the lab with the parts for their manufacturing process and would not be able to spare a large number of parts for testing. In addition, running the testing with parts would not allow for quick analytical assessment of the various alternative cleaners. Coupons allow for a known amount of soil to be added and provide for an easy way to measure removal rates. They take part geometry out of the equation and allow the focus to be on the chemical attributes of the alternatives being analyzed and tested. Once the working number of possible alternative products is reduced to a smaller number, only a few actual parts would be needed to conduct the performance testing comparison. Figure 7-4 shows a soiled part ready for cleaning trials.

IV. Actual Product Cleaning Studies
Geometries and sizes of parts important to cleaning efficiency

Figure 7-4. Phase IV Part to be Cleaned.

Testing will follow the developed methodology from the previous phases along with input from the end user and final success or failure of the lab evaluation will be made by both the TURI Cleaning Lab and the end users. This is done because the end user should be considered the expert in the specific process. In the past, this was where the project ended, with the lab sending the cleaned parts back to the company, providing the end user with contact information for the two or three vendors of the successful products. It was then left up to the end user to complete the project on their own. Following this process a company would complete the project in about a third of the time [26]. Even though this seems low, the success rate for similar projects outside of the TURI Cleaning Lab was far lower. Technical assistance in pollution prevention activities has had marginal success over the years with a national average adoption rate of around 10% [27].

To improve the adoption of safer alternatives, a different approach was needed. At the TURI Cleaning Lab, a more comprehensive on-site follow up assistance program was developed and implemented. The result was the addition of the Phase V, on-site, scale-up feasibility study. The lab determined that many times the final assessment of cleaned parts could be made on location as part of the final phase of the cleaning project.

PHASE V PILOT PLANT/SCALE-UP FEASIBILITY STUDIES

With safer alternatives to TCE identified and evaluated, the TURI Cleaning Lab would then provide companies with the opportunity to have small scale equipment loaned to them (free of charge) from the TURI Cleaning Lab to evaluate the identified products on-site. This way, the company would be able to set up and run batch cleaning testing with parts that come directly off the production line. This real time evaluation of the proposed cleaning process is shown in Figure 7-5. Equipment and cleaning alternative supplies were left with a company for a period of two to four weeks. At the end of the lending period, the equipment was retrieved and the status of the alternative cleaning product was determined.

Figure 7.5. Phase V Pilot Testing.

Initially, the equipment and chemistry loaners helped more companies but it was still not enough. A critical component was missing, personnel training. The position of field scientist/engineer for the TURI Cleaning Lab staff was implemented to further assist companies with the piloting process. This position allowed the lab to make more personal connections. Assistance may include more lab specific testing, loaning of equipment to the end user to trial the alternatives, or pilot testing with the field scientist/engineer present. In addition, the lab has helped facilitate companies in working with vendors to obtain sample solvents or additional, larger equipment loaners. Investigation into end user clients' testing needs due to military, aerospace, or FDA specifications could also happen during this part of the process. The TURI Cleaning Lab staff acts as another resource and a potential project manager for the end users and has helped companies implement the process rather than by just providing them the materials to do so on their own.

In addition, the on-site process was intended to include the employees who are responsible for cleaning at the company. Employee input about the process allowed a greater opportunity for successfully adopting the laboratory recommended alternative cleaning process. Furthermore, input from the company's management is important to ensure final adoption will be financially acceptable.

Prior to the interactive, hands-on assistance, the adoption of safer alternatives has been estimated to reach around a third of the companies looking for a new cleaning process. With the expanded effort, acceptance of safer and effective products into the companies operating plans has been shown to reach nearly three quarters of those conducting on-site process specific testing of bench-scale proven products and equipment.

OTHER METHODS TO IMPROVE ADOPTION

In some industry sectors, there are situations that arise when a group of companies are doing work that is very similar and utilize many of the same operating processes. To be able to reach a sector more efficiently, hands-on workshops have been successful in getting the message out with minimal effort from the TURI Cleaning Lab.

Half-day, hands-on training workshops for similar companies have provided participants with a quick way to see that successful alternatives do exist for their cleaning needs. To maximize the time of the participants at the workshop, dirty parts and soils were collected from a sampling of the related companies prior to the event. This allowed the lab to identify potential products and pretest them to ensure they work on the soils and parts. This end-user-specific testing was deemed necessary as solvent substitution in cleaning applications does not have a single drop-in. By providing testing services to companies, the TURI Cleaning Lab was able to come to the workshop with the specific products that were most likely to be effective for individual companies.

During the workshop, companies brought their dirty parts to be cleaned using the pre-identified alternatives from the TURI Cleaning Lab. The workshop allowed these companies to see firsthand that alternatives do exist that are effective for their needs. An added benefit of the workshop was that it gave these small companies the opportunity to talk with their peers to see what others have tried or experienced. The networking provided the companies with a sense of knowing that they were not alone and emboldened them to begin the solvent replacement process. After the workshop, follow-up is provided through the on-site assistance from the TURI Cleaning Lab.

From past experience, many of the companies that have attended these industry specific workshops made the switch to a safer solvent (roughly 60%). In some cases, a few companies that received testing services from the TURI Cleaning Lab workshop did not adopt the suggested alternative cleaning products that were identified. However, the proof from peer companies that the alternatives to TCE could work prompted the companies to investigate other possible substitutes on their own (another 10%).

SUMMARY

The key to identifying a successful alternative is not relying on a silver bullet to solve the problem. Instead, a systematic methodology of performance evaluations, both in the lab and in the field, tailored to the needs of the individual project and company provides the best avenue to successfully replacing TCE. Working directly with those that will be running the cleaning line at the company is paramount. The TURI Cleaning Lab has fine-tuned its process to combine all of these steps.

Chapter 8

IF NO SILVER BULLET WHAT TO TEST

ABSTRACT

Based on the work of the TURI Lab, the alternatives to TCE are evaluated for performance and basic safety profiles. Utilizing safety screening tools and more comprehensive assessment methods will help to keep a company from making an unforeseen switch to something that was not an improvement over TCE.

The previous chapter discussed how to test products. Chapter 2 also briefly described some of the classifications of the possible alternatives to TCE. In this chapter, the performance and basic safety screening of the products tested at the TURI Cleaning lab over the past 20 years are reviewed. The TURI Cleaning Lab safety screening score (described in Chapter 6) was calculated for many of the traditional solvents and the alternatives selected by the TURI Cleaning Lab as possible substitutes. Also in this chapter the P2OASys tool discussed in Chapter 6 is used to complete an expanded review of health and safety elements. A comparison of a handful of top used aqueous products and possible drop-in vapor degreasing solvent alternatives is provided.

To begin the process, the safety screening score for TCE was calculated at 37 from the data in Table 8-1. With this as the benchmark value, the goal during testing was to meet or exceed this value for the potential alternatives. This comparison is only the first step in the larger effort to assess environmental health and safety.

Table 8-1. TCE Safety Screening Score

Indicator	Value	Points
VOC:	1470	0
GWP:	0	10
ODP:	0	10
NFPA H:	2	
NFPA F:	1	7
NFPA R:	0	
pH:	NA	10

A quick comparison shows that a wide majority of the products tested for potentially replacing TCE were alkaline aqueous. The safety screening scores for these products ranged from a low of 27 to a high of 49. Many times the low safety screening scores were included for comparative purposes or were identified by the end users as a product of interest for the lab to test. Table 8-2 lists the breakdown of product types tested and provides a range for the safety screening scores.

Table 8-2. Product Classifications for TCE Alternatives Tested at TURI Cleaning Lab

Classification	Number of records	Safety score range
Alcohol	1	42
Alcohol-Semi Aqueous	4	33
Alkaline Aqueous	426	27-49
Biobased	4	43-47
Caustic	4	37-40
Ester	12	43-47
HCFC	25	33
HFE	47	28-39
Neutral Aqueous	29	46-50
nPB	117	23-29
Organic	9	38-46
Petroleum Distillate	30	37-48
Semi-Aqueous	4	37
Terpene	26	37-48
Volatile Methyl Siloxanes	9	36-42

A portion of the testing conducted at the lab was to evaluate potential drop-in solvents (requiring little to no change in equipment or operating conditions) for vapor degreasing. During this time period, many solvents with lower safety screening scores than TCE were included in the testing. At the time, one classification of product, n-propyl bormide (nPB), was evaluated due to the solvent not being regulated. It performed similarly to TCE and was only marginally more expensive. The higher cost of the solvent was nearly eliminated when looking at the savings from the lower operating temperature of nPB as compared to TCE. In addition, companies were using nearly 25% less solvent to complete the same level of work. Therefore, the company were buying less new solvent and disposing of less waste solvent resulting in making the switch from TCE to nPB cost effective. These characteristics made the solvent very appealing to many end users looking to replace TCE. Despite the lack of reduction of hazardous level of the solvent, the end users were still willing to switch to avoid the existing regulations (or lack thereof). A decade later, the regulatory nature has shifted to include nPB products (due to their now recognized toxicity) and these end users are currently going through the replacement process again.

In all, eighty products have been evaluated specifically for TCE replacement. Some of the products are no longer available; some of the vendors have changed names or have gone out of business; but all products have been included to help illustrate what types of products have been used over the years. The top four products most often used were alkaline aqueous products. Table 8-3 lists all of the products and their safety scores and usage levels.

Table 8-3. Products Evaluated for TCE Replacement

Usage Count	Company	Products	Safety Screen Score	Classification
93	Oakite Products	Inproclean 3800	42	Alkaline Aqueous
51	AW Chesterton	KPC 820 N	46	Alkaline Aqueous
45	US Polychem Corporation	Polyspray Jet 790 XS	45	Alkaline Aqueous
32	Alconox Inc	Detergent 8	40	Alkaline Aqueous
31	Enviro Tech International Inc	Ensolv	24	nPB
25	AGA Chemical	AK 225	33	HCFC
25	Gemtek Products	SC Aircraft and Metal Cleaner	49	Alkaline Aqueous
24	Magnaflux	Daraclean 283	38	Alkaline Aqueous
24	Petroferm Inc	Lenium CP	29	nPB
22	DuPont	Vertrel MCA	28	HFE

Table 8-3. (Continued)

Usage Count	Company	Products	Safety Screen Score	Classification
18	Brulin Corporation	Aquavantage 1400	46	Alkaline Aqueous
17	Bio Chem Systems	Bio T Max	37	Terpene-Semi-Aqueous
17	Warren Chemical Company	Sea Wash Blue	50	Neutral Aqueous
16	Oakite Products	Inproclean 2000	45	Alkaline Aqueous
16	Kyzen Corporation	Metalnox M6960	23	nPB
15	Calgon Corporation	Geo Guard 2215	27	Alkaline Aqueous
14	Poly Systems USA Inc	Solvon PB	24	nPB
13	Today & Beyond	Beyond 2001	38	Alkaline Aqueous
13	Bio Chem Systems	Green Stuff (6325)	43	Alkaline Aqueous
13	Matchless Metal Polish Company	MC 132	39	Alkaline Aqueous
11	Petroferm Inc	Lenium ES	29	nPB
11	DuPont	Vertrel CCA	28	HFE
10	AW Chesterton	278 Super Solv	37	Petroleum Distillate
10	Poly Systems USA Inc	Solvon IP	29	nPB
10	International Products Corporation	Surface Cleanse 930	46	Neutral Aqueous
9	Enviro Tech International Inc	Ensolv A	24	nPB
9	US Polychem Corporation	Polyspray Jet 790 P	46	Alkaline Aqueous
8	Solvent Kleene Inc	D Greeze 500 LO	38	Organic
7	Magnaflux	Daraclean 282 GF	38	Alkaline Aqueous
7	Dysol	DS 108	37	Petroleum Distillate
7	Metabolix Inc	Metabolix E3HB	43	Ester
6	International Products Corporation	Micro 90	46	Alkaline Aqueous
6	Bio Chem Systems	Solsafe 245	37	Petroleum Distillate
5	Oakite Products	Inproclean 4000 T	40	Terpene-Organic
5	Matchless Metal Polish Company	MC 580	39	Alkaline Aqueous
4	Magnaflux	Daraclean 232	40	Alkaline Aqueous
4	Magnaflux	Daraclean 282	36	Alkaline Aqueous
4	Dysol	DS 104	37	Petroleum Distillate
4	3M	HFE 7100	39	HFE
4	Twin Rivers Technologies	Methyl Ester 1618	45	Ester
4	Kyzen Corporation	Optisolv OP7168	33	Alcohol Semi Aqueous

Usage Count	Company	Products	Safety Screen Score	Classification
4	Hubbard Hall Inc	Ultrasoak 127	37	Alkaline Aqueous
4	Ecolink	Vortex	37	Semi-Aqueous
3	Today & Beyond	Beyond 2008	48	Terpene-Semi-Aqueous
3	Solvent Kleene Inc	D Greeze 1000	48	Petroleum Distillate
3	Micro Care	Flux Remover C	33	HFE
3	Micro Care	Heavy Duty Degreaser C	33	HFE
3	Mirachem Corporation	Mirachem 500	42	Alkaline Aqueous
3	Dow Chemical Company	OS 10	36	Volatile Methyl Siloxanes
3	Dow Chemical Company	OS 20	36	Volatile Methyl Siloxanes
3	Dow Chemical Company	OS 30	42	Volatile Methyl Siloxanes
3	Buckeye International	Shopmaster HP	45	Alkaline Aqueous
3	Chrisal USA Inc	Super CMF 240	39	Alkaline Aqueous
3	Buckeye International	XL 100	41	Alkaline Aqueous
2	Today & Beyond	Beyond 2003	38	Alkaline Aqueous
2	Brulin Corporation	Compliance	37	Caustic
2	Fine Organic Corporation	FO 2085 M	41	Alkaline Aqueous
2	Brulin Corporation	Formula 815 GD	45	Alkaline Aqueous
2	3M	HFE 7200	31	HFE
2	Petroferm Inc	Lenium GS	29	nPB
2	Kyzen Corporation	Metalnox M6310	43	Alkaline Aqueous
2	US Polychem Corporation	Polychem DEOX 007	45	Alkaline Aqueous
2	Gemtek Products	SC Supersolve	47	Biobased
2	Buckeye International	Shopmaster LPH	48	Alkaline Aqueous
2	Texo Corporation	Texolite 1734 XL	47	Alkaline Aqueous
1	Ardrox Inc	6333	40	Caustic
1	Bio Chem Systems	Bio T 300 B	37	Terpene-Semi-Aqueous
1	Newport Biodiesel	Biodiesel	43	Biobased
1	Lifetime Solutins Ltd	Commercial All Purpose	48	Alkaline Aqueous
1	Magnaflux	Daraguard 416	46	Organic
1	Man Gill Chemical Company	Gillite 0650 Cl	35	Alkaline Aqueous
1	3M	HFE 71DE	39	HFE
1	Kyzen Corporation	Ionox HC 2	42	Alcohol
1	MacDermid Industrial Products	ND 17	39	Alkaline Aqueous
1	US Polychem Corporation	Polyspray Jet 790 C	39	Caustic

Table 8-3. (Continued)

Usage Count	Company	Products	Safety Screen Score	Classification
1	Warren Chemical Company	Sea Wash 8 New	50	Neutral Aqueous
1	AG Environmental Products	Soy Clear 1500	47	Ester
1	AG Environmental Products	Soy Gold 2000	47	Biobased
1	Water	Steam	50	Neutral Aqueous
1	DuPont	Vertrel XP 10	38	HFE

Table 8-4. Top Cleaning Product Comparison

Indicator	Inproclean 3800	KPC 820 N	Polyspray Jet 790 XS	Detergent 8
VOC:	0	23	0	200
GWP:	0	0	0	0
ODP:	0	0	0	0
NFPA H:	2	1	1	0
NFPA F:	0	0	0	0
NFPA R:	0	1	0	0
pH:	12	9.8	11.5	11
Safety Screen Score	42	46	45	40
Number of Trials	353	71	127	221
Effective Tests	253 (71%)	55 (77%)	83 (65%)	158 (71%)
Soils Tested	27	7	23	13
Substrates Cleaned	19	9	10	13
Equipment Used	6	4	4	4
Major Ingredients	*Sodium metasilicate *Nonyl phenolxy polyethoxy ethanol *Sodium xylenesulfonate	*Carbonic acid, sodium salt (by-products of sodium bicarbonate) *Ethylene diamine tetra acetic acid, tetrasodium salt	*Sodium silicate *Tetra potassium pyrophosphate	*Sodium dodecyl benzene sulfonate *Tetrasodium pyrophosphate *Pentasodium triphosphate *Sodium carbonate

A quick review of the material safety data sheets for each of the four top successful products reveals that no listed ingredient appears in all of the products. Using the CleanerSolutions comparison feature, the overall cleaning profiles for the products (beyond just TCE replacement trials) can be performed. The Inproclean 3800 was by far the most widely used product on more soils, substrates, and equipment combinations. Table 8-4 lists the four aqueous products and how they have been tested in the TURI Cleaning Lab.

Of the twenty drop-in solvents evaluated at the TURI Cleaning Lab, eight were n-propyl bromide based; eight were hydrofluoroethers (HFE); three were volatile methyl siloxanes; and the last was a hydrochlorfluorocarbon (HCFC). Table 8-5 lists an example product from each of the product classifications. Safety screening scores were far lower than the aqueous based products, ranging from a low of 24 to a high of 42.

Table 8-5. Drop-In Solvent Comparison

Indicator	Ensolv	AK 225	Vertrel MCA	OS 10
VOC:	1333	0	536	0
GWP:	0.0005	> 1	806	low
ODP:	0.016	> 1	0	0
NFPA H:	1	1	1	1
NFPA F:	0	1	0	3
NFPA R:	0	0	1	0
pH:	NA	NA	NA	NA
Safety Screen Score	24	33	28	36
Number of Trials	47	57	50	22
Effective Tests	45 (96%)	45 (79%)	42 (84%)	18 (82%)
Soils Tested	7	8	6	4
Substrates Cleaned	6	9	7	3
Equipment Used	3	5	3	3
Product Classification	n-PB	HCFC	HFE	VMS

Each category of the drop-in solvents was found to be effective on most of the soils tested based on the testing at the TURI Cleaning Lab. The least effective drop-in solvent was found to be effective for more than 79% of the trials. In comparison, the best aqueous based product was found to be effective in 77% of the trials. For the aqueous based products, the lower effectiveness on soil removal illustrates that there is no silver bullet to be found for these types of products. Another disadvantage of the aqueous based products is that their use does not offer the chief advantage of the vapor degreasing process, which is that the last fluid contact of the part being cleaned is with pristine solvent. Water-based products require additional stages in the cleaning process to address this shortcoming. Rinsing (typically requiring multiple stages) and drying processes are needed and add to the cost of cleaning operations.

EXPANDED EHS COMPARISON - P2OASYS

As discussed, P2OASys allows the user to assess the potential environmental, worker, and public health impacts of alternative technologies. It builds upon the TURI Cleaning Lab's safety screening methods to give a broader review of the products being considered. Typically, this assessment is completed on products that have gone through the lab screening and subsequent performance testing but prior to taking a product out for field evaluations. This is to reduce the amount of time spent gathering data on a product prior to cleaning trials and to ensure no additional safety concerns exist with the product before promoting them as a safer alternative.

Of the possible 61 data points to collect for the evaluation, the eight alternatives and TCE averaged 36 entries. The comprehensive evaluation had the KPC 820 N as the safest of the nine products (as did the safety screening scores. The most hazardous of the products was TCE (where the screening had Ensolv as the lowest). Table 8-6 lists the summary P2OASys scores and the number of data points collected for the evaluated cleaners.

Table 8-6. P2OASys Evaluation of Alternatives to TCE

Category	Inproclean 3800	KPC 820 N	Polyspray Jet 790 XS	Detergent 8	Ensolv	AK 225	Vertrel MCA	OS 10	TCE
Acute Human Effects	6	4	10	5	6	5	8	6	9
Chronic Human Effects	4	2	2	2	6		4		7
Physical Hazards	2	2	2	2	9	6	6	7	7
Aquatic Hazard				4	2		6		7
Persistence/Bioaccumulation		3	2		6				9
Atmospheric Hazard	2	2	2	2	10	10	6		6
Disposal Hazard	5	2		4	6	5			8
Chemical Hazard	8	8	7	7	10	8	10	10	10
Energy/Resource Use	6	6	6	6	8	7	8	5	8
Product Hazard	6	2	6	6	6	6	6	8	7
Exposure Potential	6	6	6	6	6	6	6	6	8
Final	45	37	43	44	75	53	60	42	86
Weighted Final (10 = worst)	5.00	3.70	4.78	4.40	6.82	6.63	6.67	7.00	7.82
Categories Used	31	39	36	30	46	34	39	21	48

The safety screening scores usually take about five minutes to gather and assess from a safety data sheet, whereas with P2OASys you can spend considerable more time collecting and analyzing the data. In the end, the screening methodology does a good job to identify the safer options. Ideally, the P2OAsys comparison should include all of the 61 data points if possible. However, realistically this may not be possible due to the lack of certain kinds of data but the tool is designed to be able to adjust the final scores based on the total number of points collected.

SUMMARY

With thousands of possible alternative products to select for many cleaning uses of TCE, a few tools can be used to help screen products out. Using the TURI Cleaning Lab's safety screening methodology is a quick and reliable way to identify potentially safer alternatives. With a shorter list to work from, selecting possible candidates becomes easier. Performance testing can be conducted to determine which of the safer products work the best. Of the more effective and potentially safer products, the list can be narrowed down further through the more comprehensive health and safety assessment. The lab proven products that have been reviewed for safety can then make their way into pilot testing with the end user.

Chapter 9

IF NO SILVER BULLET WHERE ELSE TO LOOK

ABSTRACT

Several non-profit and government sponsored programs exist to help a company find TCE replacements. It's important to note that a green cleaning program goes beyond chemical and equipment choices. It includes policies, procedures, trainings and shared responsibility efforts that minimize the impact of cleaning materials on the health of building occupants and protects the environment and the community they do business in as a whole. A review of a few of the programs and their requirements for being considered a safer choice is provided.

A green cleaning program goes beyond chemical and equipment choices. It includes policies, procedures, trainings, and shared responsibility efforts that minimize the impact of cleaning materials on the health of building occupants and protect the environment and the community they do business in as a whole.

As mentioned before, performance is not the only selection criterion that needs to be addressed when looking for a new cleaning product. Traditional solvents have known environmental, health, and safety (EH&S) effects, which has led to the development of alternative chemical formulations. The toxicity, flammability, and smog producing effects of petroleum-based solvents were recognized from an early date to be an area of concern.

The most important consideration when choosing an alternative is to not shift the risk involved in the process. For example, when attempting to replace a flammable solvent, it would not be advantageous to select an ozone depleting chemical. In this case, the risk goes from the worker (burn, explosion, etc.) to the environment (ozone destruction). Ideally, a selection of an alternative cleaning product should be safer for one or the other or if possible, a product that would result in a reduced risk for both the worker and the environment.

Discussed in previous chapters, there are several tools that can assist in the assessment of a product's safety. These tools can be helpful to someone who has the time and knowledge to use and understand the results obtained from the assessment process. However, a majority of people do not have the time to learn, use, and implement such systems. For those types of organizations/ companies, all is not lost.

There are a few third party recognition/certifying groups that can provide a reliable list of products that have met a certain level of EHS review and performance levels that would meet the needs of the industrial cleaning scenarios. These include EPA's Safer Choice program, Green Seal, and United Lab ECOLOGO. CleanGredients is an additional resource available online that can help in assessing the safety of various products or ingredients used in the cleaning industry (although mostly for janitorial and household applications).

EPA Safer Choice Program

The Environmental Protection Agency's Design for the Environment (now known as the

Safer Choice Program) was developed to work with a wide range of stakeholders to help reduce the risk of chemical hazards to people and the environment. The reduction effort works toward preventing pollution by focusing on industries that combine the potential for chemical risk reduction and improvements in energy efficiency with a strong motivation to make lasting, positive changes.

The stakeholders include representatives from these industries as well as environmental groups. Working together, they have been able to develop goals and guidance for the work of the group. Partnership projects evaluate human health and environmental characteristics, performance, and cost of traditional and alternative technologies, materials, and processes. As incentives for participation and driving change, Safer Choice Program offers unique technical tools, methodologies, expertise, and the potential for product recognition [28].

The Safer Choice Program focuses on promoting safer product development and using green chemistry alternatives through researched substitutions. The goal of this type of substitution is to reduce unintended consequences. To be considered safer choices, potential alternatives typically exhibit as many of the following characteristics as possible: they should be technically feasible; deliver the same or better value in cost; deliver the same or better value in performance; provide an improved profile for health and the environment; account for economic and social considerations; and have the potential to result in lasting change [29].

The Safer Choice Program Standard for Cleaning Products is intended to cover a wide range of products, including but not limited to: glass cleaners, general purpose cleaners, washroom cleaners, carpet cleaners, laundry detergents, graffiti removers, boat and car cleaning, drain cleaners, personal care, and floor care and other industrial products.

To be able to cover a wide range of product types, the Safer Choice Program Master Criteria for Safer Ingredients (Master Criteria) was developed. It includes comprehensive, science-based criteria designed to ensure that the safest possible ingredients are used in the various Safer Choice Program-labeled products. Safer Choice Program looks at the ingredients of a product and not just the end product itself. This evaluation looks at an ingredient within its functional class context and on the chemical's key, distinguishing human health and environmental characteristics. By doing this, potential product ingredients can be viewed as part of a continuum of improved or safer ingredient choices.

The Master Criteria make it possible to set the spectrum for the greener or "low-concern" end of the continuum of chemical safety. To define low concern, Safer Choice Program uses toxicological levels established by respected health and environmental protection authorities, including the United Nation's Globally Harmonized System (GHS) for the Classification and Labeling of Hazard Substances and the U.S. EPA's New Chemicals Program. For functional classes where no low-concern ingredients currently exist, Safer Choice Program works with its stakeholders to carefully modify the Master Criteria in a way that allows for ingredient choices while ensuring the safest possible ingredients in that functional class. These criteria were designed for use in distinguishing safer chemicals for the Safer Choice Program and are listed in Table 9-1.

Table 9-1. Safer Choice Program Master Criteria

Category	Thresholds	
Acute Mammalian Toxicity	Route (units)	Median Lethal Dose/Concentration
	Oral, LD50 (mg/kg-bw) (by weight)	2,000
	Dermal, LD50 (mg/kg-bw)	2,000
	Inhalation, LC50 (vapor/gas) (mg/L)	20
	Inhalation, LC50 (dust/mist/fumes) (mg/L)	5
	R20: Harmful by inhalation R21: Harmful in contact with skin R22: Harmful if swallowed R23: Toxic by inhalation R24: Toxic in contact with skin R25: Toxic if swallowed R26: Very toxic by inhalation R27: Very toxic in contact with skin R28: Very toxic if swallowed *And all combination risk phrases containing one or more of the above (R20-R28).*	
	H300: Fatal if swallowed H301: Toxic if swallowed H302: Harmful if swallowed H310: Fatal in contact with skin H311: Toxic in contact with skin H312: Harmful in contact with skin H330: Fatal if inhaled H331: Toxic if inhaled H332: Harmful if inhaled	

Category	Thresholds	
Carcinogenicity	National Toxicology Program	Known to be Human Carcinogen Reasonably Anticipated to be Human Carcinogen
	EPA	(2005/1999) Carcinogenic to humans, Likely to be carcinogenic to humans, or Suggestive evidence of carcinogenic potential -1 (1996) Known/Likely (1986) Group A – Human Carcinogen, Group B – Probable human carcinogen, or Group C – Possible human carcinogen -1
	IARC	Group 1 – Carcinogenic to humans Group 2A – Probably carcinogenic to humans Group 2B – Possibly carcinogenic to humans -1
	European Union carcinogenic, mutagenic and toxic to reproduction substances (EU CMR List)	Category 1 – Known to be carcinogenic to humans Category 2 – Should be regarded as if carcinogenic to humans Category 3 – Cause for concern for humans owing to possible carcinogenic effects
	EU Risk Phrases	R45: May cause cancer R49: May cause cancer by inhalation R40: Limited evidence of a carcinogenic effect And all combination risk phrases containing R45, R49, or R40
	EU Classification, Labeling, and Packaging (CLP)	H350: May cause cancer H350i: May cause cancer by inhalation H351: Suspected of causing cancer
	NIOSH Occupational Carcinogen List	http://www.cdc.gov/niosh/topics/cancer/npotocca.html
	Globally Harmonized System (GHS)	Category 1A – Known to have carcinogenic potential for humans Category 1B – Presumed to have carcinogenic potential for humans Category 2 – Suspected human carcinogens

Table 9-1. (Continued)

Category	Thresholds	
Genetic Toxicity	EU CMR List	Category 1 – Substances known to be mutagenic to man Category 2 – Substances which should be regarded as if they are mutagenic to man Category 3 – Substances which cause concern for man owing to possible mutagenic effects
	EU Risk Phrases	R46: May cause heritable genetic damage R68: Possible risk of irreversible effects *And all combination risk phrases containing R46 or R68.*
	CLP	H340: May cause genetic defects H341: Suspected of causing genetic defects
	GHS	Category 1A – Chemicals known to induce heritable mutations in germ cells of humans Category 1B – Chemicals which should be regarded as if they induce heritable mutations in the germ cells of humans Category 2 – Chemicals which cause concern for humans owing to the possibility that they may induce heritable mutations in the germ cells of humans
Neurotoxicity	Route (units)	Median Lethal Dose/Concentration
	Oral (mg/kg-bw/day)	100
	Dermal (mg/kg-bw/day)	200
	Inhalation (vapor/gas) (mg/L/6h/day)	1
	Inhalation (dust/mist) (mg/L/6h/day)	0.2
Repeated Dose Toxicity	Oral (mg/kg-bw/day)	100
	Dermal (mg/kg-bw/day)	200
	Inhalation (vapor/gas) (mg/L/6h/day)	1
	Inhalation (dust/mist) (mg/L/6h/day)	0.2
	EU Risk Phrases	R48: Danger of serious damage to health by prolonged exposure (repeated exposure) And all combination risk phrases containing R48.
	CLP	H372: Causes damage to organs H373: May cause damage to organs

Category	Thresholds	
	Route (units)	Median Lethal Dose/Concentration
	Oral (mg/kg-bw/day)	250
	Dermal (mg/kg-bw/day)	500
	Inhalation (vapor/gas) (mg/L/6h/day)	2.5
	Inhalation (dust/mist) (mg/L/6h/day)	0.5
Reproductive and Developmental Toxicity	EU CMR List -3	Category 1 –Substances known to impair fertility in humans or known to cause developmental toxicity in humans Category 2 – Substances which should be regarded as if they impair fertility in humans or cause developmental toxicity to humans Category 3 – Cause concern for human fertility or possible developmental toxic effects
	EU Risk Phrases -3	R60: May impair fertility R61: May cause harm to the unborn child R62: Possible risk of impaired fertility R63: Possible risk of harm to the unborn child R64: May cause harm to breastfed babies R33: Danger of cumulative effects And all combination risk phrases containing R60-R64 or R33
	CLP -3	H360: May damage fertility or the unborn child H361: Suspected of damaging fertility or the unborn child H362: May cause harm to breast-fed children
Respiratory Sensitization	EU Risk Phrase	R42: May cause sensitization by inhalation
	CLP	H334: May cause allergy or asthma symptoms or breathing difficulties if inhaled
	GHS	Category 1A – high frequency of occurrence or sensitization rate in humans Category 1B – low to moderate frequency of occurrence or sensitization rate in humans

Table 9.1. (Continued)

Category		Thresholds		
Skin Sensitization		EU Risk Phrase	R43: May cause sensitization by skin contact	
		CLP	H317: May cause an allergic skin reaction	
		GHS	Category 1A – high frequency of occurrence in humans and/or a high potency in animals Category 1B – low to moderate frequency of occurrence in humans and/or a low to moderate potency in animals	
Environmental Toxicity and Fate	If a chemical is an acute aquatic toxicant (i.e., L/E/IC50 < 100 ppm), then it must biodegrade rapidly and not be bioaccumulative (see lines 1-3 below). If a chemical has low aquatic toxicity (see line 4 below), then its half-life must be less than 60 days.			
	Line	Acute Aquatic Toxicity Value (L/E/IC50) 4,5,6	Persistence (Measured in terms of level of biodegradation)	Bioaccumulation Potential
	1	If ≤ 1 ppm…	…then may be acceptable if the chemical meets the 10-day window as measured in a ready biodegradation test without degradation products of concern8…	…and BCF/BAF < 1000.
	2	If > 1 ppm and ≤ 10 ppm…	…then the chemical must meet the 10-day window as measured in a ready biodegradation test without degradation products of concern …8	
	3	If > 10 ppm and < 100 ppm…	…then the chemical must reach the pass level within 28 days as measured in a ready biodegradation test without degradation products of concern8…	
	4	If ≥100 ppm…	…then the chemical need not reach the pass level within 28 days as measured in a ready biodegradation test if there are no degradation products of concern8 and its half-life < 60 days…	

Eutrophication	The total level of phosphorus in the cleaning product will be limited to a maximum level of 0.5 weight % in the cleaning product as sold (measured as elemental phosphorus). Inorganic phosphates, as defined by the US EPA New Chemicals Program, cannot make up any portion of the 0.5 weight % of phosphorus.

1 - Chemicals listed as "possibly carcinogenic to humans" are evaluated largely on animal studies. Safer Choice Program will consider appropriate data that show cancer concerns are not relevant to humans, e.g., because of an animal specific tissue effect or mode of action. If the data demonstrate that cancer concerns are not relevant to humans, that chemical can be considered under the Safer Choice Program Criteria.

2 Per EU guidance, chemicals classified as Category 3 substances may be placed in that category based on positive results in assays showing (a) mutagenic effects or (b) other cellular interaction relevant to mutagenicity. If a chemical is classified in Category 3(b) only, and that classification appears overly conservative, then the submitter may request EPA expert review. In such as case, if EPA determines the data do not support a concern for possible mutagenic effects, then the chemical will pass the criteria.

3 The EU classification criteria do not currently consider a limit dose above which an adverse effect would not trigger classification. EPA will consider evidence demonstrating that a chemical carrying a reproductive/developmental toxicity risk phrase or listed as toxic to reproduction (in Table 6b) did not cause an adverse effect below the TSCA 8(e) Guidance Values listed in Table 6a. Such a chemical may be determined, upon EPA review, to pass the DfE criteria for reproductive/developmental toxicity.

4 In general, there is a predictable relationship between acute aquatic toxicity and chronic aquatic toxicity for organic chemicals, i.e., chemicals that have high acute aquatic toxicity may also have high chronic aquatic toxicity at low concentrations [Rand, G.M., ed. Fundamentals of Aquatic Toxicology. 2nd ed. 1995, Taylor & Francis: Washington, DC.]. Since acute aquatic toxicity data are more readily available, the Safer Choice Program Screens use these data to screen chemicals that may be toxic to aquatic life. Where measured chronic toxicity data is available, it will be assessed with other data and applied in the screen based on the relationship between acute and chronic aquatic toxicity.

5 A case-by-case approach focusing on rate of biodegradation and degradation products of concern will be implemented for chemicals toxic to aquatic organisms at ≤ 1ppm.

6 For determining the aquatic toxicity of substances that are not toxic at their solubility limit, see ECOSAR Technical Reference Manual Figure 9, p 17 (http://www.epa.gov/oppt/newchems/tools/ecosartechfinal.pdf); When a chemical may have effects at saturation as determined using the guidance in the ECOSAR manual, a weight-of-evidence approach in combination with US EPA expert review will be used. EPA may require additional testing, including, but not limited to, solubility testing, chronic aquatic toxicity testing, or acute aquatic toxicity testing of analogs.

7 Degradation products of concern are compounds with high acute aquatic toxicity (L/E/IC50 ≤ 10ppm) which mineralize < 60% in 28 days.

Green Seal

Green Seal uses science-based programs to empower consumers, purchasers, and companies to create a more sustainable world. The most applicable standard for TCE substitution would be the Environmental Standard for Cleaning and Degreasing Agents (GS 34). This standard contains the basic requirements for cleaner/degreasers suitable for cleaning soils in production and maintenance applications. The requirements are based on an assessment of the environmental impacts of product manufacture and use, and were developed with the guidance and input of a stakeholder committee consisting of solvent manufacturers, solvent users, government officials, academic and consulting experts, and environmentalists. The standard was reviewed and approved by consensus of the stakeholder committee, as defined by the International Organization for Standardization and the Federal Government (Office of Management and Budget Circular A-119) [30].

For performance evaluations, the cleaning/degreasing agent shall clean a steel coupon to a specified cleanliness level for two types of soil specified in the test method. The soils are meant to represent two types of contamination. The first is a maintenance soil that includes carbon black, iron oxide, WD-40, AW32 Hydraulic Oil, and Hypoid SAE 140 Gear Oil. The second is a production soil and includes quench oil and honing/Cutting Oil. The standard does state that end users may need to conduct their own performance testing to determine if a degreasing agent meets specific cleaning requirements. The EHS criteria for Green Seal standard GS 34 is listed in Table 9-2.

Table 9-2. Green Seal GS 34 Criteria

Category	Thresholds	
Toxic Compounds	Oral LD50	< 5,000 mg/kg
	Inhalation LC50	< 20,000 ppm of vapor or gas
		500 mg/L of mist, dust, or fumes
	Dermal LD50	< 2,000 mg/kg
Carcinogens and Reproductive Toxins	International Agency for Research on Cancer	Group 1 carcinogenic to humans Group 2A probably carcinogenic to humans Group 2B possibly carcinogenic to humans agents
	State of California under the Safe Drinking Water and Toxic Enforcement Act of 1986	Chemicals known to cause reproductive toxicity
Corrosivity and Causticity	pH of the degreasing agent concentrate shall be	< 11.0 but > 2.5

Category	Thresholds	Category
Skin and Eye Irritation	Shall not be a skin or eye irritant	Considered a skin irritant if it has a mean score of 2 or more for either erythema and eschar formation or edema formation, based on the OECD dermal scoring system (OECD, TG 404).
		Eye irritant if it causes significant ocular lesions in any type of ocular tissue (i.e., cornea, iris, or conjunctivae) within 72 hours after exposure that persist for at least 24 hours.
		Product concentrate shall meet all of the following criteria: Mean score for cornea opacity of < 2 Mean score for iris lesions of < 1 Mean score for redness of the conjunctivae of < 2.5 Mean score for edema of the conjunctivae (chemosis) of < 2
Flammability and Ignitability	Concentrated product shall not be ignitable	Flashpoint for the compound is > 140° F
Volatile Organic Compound (VOC) Content	VOCs include all organic compounds with vapor pressure of greater than 0.1 mm mercury at 1 atm pressure and 20° C	VOC content of the product as used shall not exceed the lower of the following options: 5% by weight The current regulatory limits of the Air Resources Board for the State of California (CARB) for its product category.
Ozone Depletion	Product shall not contain any ozone-depleting substances	
Toxicity to Aquatic Life	Compound is considered not toxic to aquatic life if it meets one of the following criteria	Acute LC50 daphnia or fish > 100 mg/L Acute LC50 algae > 100 mg/L
Aquatic Biodegradability	The compound shall meet the minimum requirement of	80% biodegradation based on dissolved organic content concentration
		70% of theoretical maximum CO_2 production within 28 days
		Exception to this criterion can be made for products that are practically insoluble, i.e., solubility less than 10 mg/L.
Eutrophication	Phosphates and phosphonates, including sodium salts and potassium salts, shall not be present in the product as used in quantities above 0.5% by weight of total phosphorus.	
Disposal	Manufacturer shall either take back unused or spent products for recycling or disposal or provide the user with specific recycling and disposal instructions.	
Animal Testing	To avoid new animal testing, previous test results will be accepted as evidence of meeting a criterion.	
	Preferred methods for new testing include methods that replace, reduce, or refine animal use, particularly those recommended by the Interagency Coordinating Committee on the Validation of Alternative Methods or the European Centre for the Validation of Alternative Methods, unless indicated otherwise.	
	Other non-animal (in-vitro) test results, modeling data, data from structural analogs, and other lines of evidence may be accepted provided that the methods are peer-reviewed and applicable.	

UNITED LAB ECOLOGO

United Lab (UL) is recognized by 66% of consumers in North America and can be found on more than 22 billion products worldwide. UL's ECOLOGO is recognized or referenced in more than 350 specifications and standards, and UL Environment has public affairs and outreach teams dedicated to enhancing market adoption [31].

UL's ECOLOGO environmental product certification process uses stringent, multi-attribute, consensus-based standards to confirm that products offer reduced environmental and health-related impacts throughout their entire lifecycle. Applicable ECOLOGO product standards are currently available for cleaning chemicals. Each ECOLOGO standard sets metrics for a host of environmental criteria, which can include: materials, product performance, human health impacts, environmental impacts, waste diversion, natural resource conservation, recyclability, product stewardship, and innovation. Table 9-3 lists the categories reviewed by ECOLOGO.

Table 9-3. Ecologo EHS Criteria

ECOLOGO Basic Criteria
Toxicity for aquatic and other organisms
Greater biodegradation
Lower potential for bioaccumulation
Reduction in eutrophication of water bodies
Limited waste creation
Reduction in resource use
Flash point
Reduction in ground-level ozone formation
Have an ozone depleting potential of zero
Not be formulated or manufactured with any chemicals that are included in the International Agency for Research on Cancer (IARC) lists for proven (Group 1), probable (Group 2A), or possible (Group 2B) carcinogens
Not be formulated or manufactured with any chemicals identified as priority for research by the European Union as endocrine disruptors

The most relevant standards for solvent cleaning from ECOLOGO include CCD-146 Hardsurface Cleaners Section F Industrial cleaners, CCD-110 and Cleaning and Degreasing Compounds: Biologically-based. Industrial and commercial cleaners are used primarily for general facility cleaning and machinery/parts cleaning. The selection of a cleaner is influenced primarily by the nature of the surface to be cleaned, the nature of the soil, and the degree of cleanliness required. The biologically-based cleaning and degreasing compounds contain enzymes and/or microbial cultures that promote microbial digestion of hydrocarbons, organic contaminants, and other undesirable substances. Biologically-based products may also contain surfactants and other compounds that increase their efficacy.

Both standards refer to performance based on at least 75% cleaning efficiency as measured by test methods A5 "particulate and oily soil/vinyl tiles," or A6 "oil, carbon, black and clay/white enamel painted stainless-steel panels" in ASTM D4488-95(2001)e1 01-Jan-1995 "Standard Guide for Testing Cleaning Performance of Products Intended for Use on Resilient Flooring and Washable Walls." Although it is easy to compare such products to each other as they are tested on the exact same simulated soils, the down fall is that the soils may not be specific to the actual cleaning scenarios of the potential end user. Process specific testing would still be needed.

For the above certifying bodies, the end user is provided with a listing of products that meet a specified level of environmental and health criteria. The listed products will also have been tested under similar conditions to provide a level of performance. However, each has their limitations in that the performance testing is based on a generic soil type and under basic operating conditions. An additional problem is that the lists will only contain products that have paid to have their products evaluated and approved. Although this provides the end user with a substantial list of products, it only includes those that have participated in the certification process. In some cases, better and safer products exist but the companies making them may be small and cannot afford the costs of some certifying programs.

CleanGredients

From GreenBlue, this online database of cleaning product ingredient chemicals provides verified information about the environmental and human health attributes of listed ingredients. GreenBlue® is a nonprofit that equips businesses with the science and resources to make products more sustainable. CleanGredients can be seen as a "one-stop shop" for green chemicals that reduces research, development, and marketing costs. For suppliers, it provides a centralized marketing venue for products. For formulators, it reduces time spent searching for appropriate ingredients, significantly increasing speed-to-market [32].

The site contains various components of cleaning products. These include chelating agents, fragrances, solvents, and surfactants. There are four levels or tiers of information tied to the various groups. The first tier requires the submission of test data for third-party review. Tier 1 attributes are considered to be key attributes for determining whether or not an ingredient will support environmentally preferable product formulation. In the second level, attributes are required, if known, and are self-reported. Attributes are also important for characterizing ingredients for environmentally preferable product formulation but do not require the submission of test data for third-party review. Next, the third tier attributes are voluntary but still environmentally relevant. The attributes are intended to provide a place for information that supports broader environmental perspectives such as life cycle assessments, risk assessments, or information on other environmentally relevant product features. The final level of data falls into the fourth tier. These are criteria that were considered by the relevant technical advisory committee but were not recommended for inclusion in CleanGredients at this time.

The two major categories of interest are solvents and surfactants. In order to be listed in CleanGredients, solvents must pass a minimal set of requirements for fate, human health, and environmental toxicity. There is a direct connection with the solvent EHS criteria with the Safer Choice Program Solvent Screen process. Unfortunately, the solvents that can be listed include alcohols, esters, ethylene glycol ethers, and propylene glycol ethers. There is an effort from Safer Choice Program to expand the list to include terpenes, amines, and amides.

Surfactants listed in CleanGredients must also meet certain minimal environmental performance requirements. Again the surfactants are tied to the relevant Safer Choice Program criteria. Both the solvent and surfactant criteria are listed in Table 9-4 for Tier 1, 2, and 3.

Table 9-4. CleanGredients EHS Criteria for Solvents and Surfactants [32]

	Tier 1 - Third-party review required	Tier 2 - Required if known	Tier 3 - Optional
Solvents	Acute Aquatic Toxicity Acute Mammalian Toxicity Bioconcentration Factor Biodegradation Carcinogenicity CMRs and PBTs	California Proposition 65 Canadian Environmental Protection Act Toxic Substances List CERCLA Hazardous Substances List Clean Air Act Hazardous Air	Aniline Point Aqueous Solubility Azeotrope Bioaccumulation Potential Boiling Point Canadian Domestic

	Tier 1 - Third-party review required	Tier 2 - Required if known	Tier 3 - Optional
	Developmental Toxicity DfE Criteria DfE Criteria for Direct Release Products DfE Maximum Use Level Exclusions Full Green Circle Half-Green Circle Neurotoxicity Repeated Exposure Systemic Toxicity Reproductive Toxicity Yellow Triangle	Pollutant List Clean Water Act Priority Pollutants List Degradation Products of Concern Dermal Absorption Endocrine Disruption EPA Inerts List Eye Irritation/Serious Eye Damage Flash Point Generally Regarded as Safe (GRAS) Genotoxicity Maximum Incremental Reactivity (MIR) Ozone-Depleting Substances SARA/EPCRA 313 List (Toxics Release Inventory) Sensitization - Respiratory Sensitization - Skin Skin Irritation/Corrosion Vapor Pressure VOCs	Substances List Chronic Aquatic Toxicity Color Distillation Range Evaporation Rate Freezing Point Kauri-Butanol Number Life Cycle Assessment Materials Compatibility Octanol-Water Partition Coefficient Origin of Feedstock Persistence pH Risk Assessment Specific Gravity Surface Tension TSCA Inventory Vapor Density Viscosity
Surfactants	Acute Aquatic Toxicity Biodegradation CMRs and PBTs Degradation Products of Concern DfE Criteria DfE Criteria for Direct Release Products DfE Maximum Use Level Exclusions Full Green Circle Half-Green Circle Yellow Triangle	% Active(s) Acute Mammalian Toxicity Alkylphenol Ethoxylates (APEs) Cloud Point Critical Micelle Concentration Density or Specific Gravity Flash Point Hydrophile-Lipophile Balance Irritancy pH Physical Form Sensitization Surface Tension VOC Content	Additional Aquatic Toxicity Endocrine Disruption EU Detergent Directive Life Cycle Assessment Origin of Feedstock Other Product Features Risk Assessment

The CleanGredients online tool provides a way for end users to educate themselves about what they should be looking for when evaluating potential cleaning products. The tool provides scientifically credible information about ingredient chemicals, including general product and regulatory information, physical-chemical properties, human and environmental health attributes, and sustainability attributes. It can help to identify preferable alternatives and provides end users with a way to contact suppliers, links to websites (including the CleanerSolutions database), material safety data sheets, technical fact sheets, and more.

To use the database most effectively, a basic knowledge of chemistry and toxicology is recommended. It is important to remember that CleanGredients is not a certification program. It is a database of chemicals suitable for use in cleaning products. The goal of CleanGredients is to help chemical formulators and suppliers develop safer, more effective products, whereas certification programs like the EPA Safer Choice Program, Green Seal, and ECOLOGO help consumers make better-informed choices.

HANSEN SOLUBILITY PARAMETERS IN PRACTICE

In several locations the Hansen Solubility parameters (HSP) have been mentioned briefly without much explanation as to what these parameters are and how to determine what they are.

HSP are tied to thermodynamics. If you want to dissolve something in something else then you need to compare two energy losses with one energy gain. The first is the interaction of the solvent with itself, the second is the interaction of the solute (what you want to dissolve) with itself. The gain is the interaction of the solvent with the solute. If this interaction is greater than the sum of the losses, then the solute will dissolve [33].

Therefore, to know if one thing dissolves in another, you would need to calculate the two losses and the one gain. For simple systems this can be done, but it becomes impossibly difficult once you start looking at more complex systems (azeotropes, for example). What HSP can do is simplify the process without compromising accuracy. The three values most important to HSP are dispersive forces, polar forces, and hydrogen bonding.

Dispersion forces are the general van der Waals interactions between just about everything. Because they are everywhere, they tend to get ignored, but they are the dominating force in most interactions.

Polar forces are the "positive attracts negative" electrical attractions arising from dipole moments. They are important in just about every molecule except some hydrocarbons and special chemicals consisting of only carbon and fluorine.

Hydrogen bond forces are a type of polar force. But their predictive value goes beyond simply thinking of them as polar forces so it is worthwhile to treat them separately. More generally, they can be considered as a form of electron exchange so that CO2 shows strong "hydrogen bonding" forces that make it a good solvent for [33].

To illustrate, when you want the HSP of a particular solvent but cannot use that solvent because it is toxic (TCE) or too expensive (HFE), how would you proceed? Select two (possibly widely different) safe/cheap solvents in proportions that give you the correct HSP and you have a fully functional solvent, indistinguishable (as far as the solute is concerned) from the original solvent.

To calculate if a product would dissolve another, Dispersion (δD), Polar (δP), and Hydrogen (δH) bonding parameters would be used in the following equation:

$$\text{Difference}^2 = 4\,(\delta DA\text{-}\delta DB)^2 + (\delta PA\text{-}\delta PB)^2 + (\delta HA\text{-}\delta HB)^2$$

The drop-in solvents are consistent in their effectiveness, but they carry the burden of a lower safety screening score, meaning likely to be unsafe to use. In addition, a majority of the drop-in solvents cost a significant amount more than TCE. Additional replacement solvents prompt fewer safety concerns but application of these solvents is limited because their solubility parameters resemble those of paraffinic solvents which do not possess significant hydrogen-bonding and polar intermolecular forces [34].

To align the hydrogen-bonding and polar forces to match those of TCE, there is a push to start to think outside the box. Instead of using a single solvent to obtain the necessary cleaning power, a combination of solvents can be used. To be able to do this in a systematic way, the use of the previously mentioned HSP would be good place to start.

Azeotropes are blends of solvents. The number of components can be two, three, four, or more. But for vapor degreasing work, more than two are probably impractical. Work has been done to evaluate several binary azeotropes for soil removal. An azeotrope is characterized by having the same composition in the liquid as in the vapor when the liquid is boiling. In this way, an azeotrope acts as if it were a single component. As a plus, binary azeotropes do not require development of new molecules, synthesis routes, manufacturing capability or, generally, regulatory acceptance.

At the TURI Cleaning Lab, in conjunction with Dr. John Durkee III, a series of performance evaluations were conducted. One solvent was chosen as a target to see if selected binary azeotropes could produce the same laboratory cleaning performance as demonstrated by the target solvent in similar laboratory cleaning experiments. Although the testing was set up to replace n-propyl bromide, the same methodology could be easily adapted to TCE. Dr. Durkee provided the lab with four binary azeotropes. These were chosen from a database containing thousands of binary azeotropes. Selection was based on the components generating a substantially lower level of health and safety than the target solvent and their expected level of cleaning performance. The selected combinations are listed in Table 9-5.

Table 9-5. Azeotrope Mixtures

Azeotrope components (wt %)	Boiling point (deg C)	VOC info	Exposure limits (ppm)
16.5 water 83.5 t-butyl acetate	87.4	Exempt in U.S.	Not listed 150
5.0 water 95 methyl acetate	56.5	Exempt in U.S.	Not listed 200
12.9 water 87.1 n-heptane	79.4		Not listed 400
51 water 49 propylene glycol methyl ether (PGME)	97.5		Not listed 100

To evaluate the effectiveness, the four supplied solvents were mixed with DI water in beakers to obtain the specified binary azeotropes. Each was heated to its boiling point. Pre-weighed aluminum coupons were coated with each of the four soils (three coupons per soil). Coupons were weighed again to determine the amount of oil applied. Three coupons were placed into the solvent vapor and cleaned. Coupons were rinsed in warm tap water and dried for 30 seconds using compressed air at room temperature. Coupons were weighed a third time to determine the amount of oil remaining. Efficiencies were calculated and recorded.

Table 9-6. Azeotrope Testing Results

Azeotrope Mixture	Castrol Quench G Oil	Cargill Canola Oil	CP Hall Co Plathall Eso Oil	Soltex Polytbutene 32 Coating
t-butyl acetate	94.62	98.37	99.50	99.40
methyl acetate	99.97	99.93	98.60	99.76
n-heptane	91.33	89.40	83.30	60.26
PGME	57.61	56.27	70.07	0.53
nPB	100.05	100.00	100.01	99.65

When looking at the results in Table 9-6, t-butyl acetate-water azeotrope and heptanes-water azeotrope performed as expected, working as well as the target solvent and outperforming the other two azeotrope blends.

This laboratory scale testing proved that using HSP is a viable way to establish another option for drop-in solvents in the search for the TCE silver bullet.

For those hard to replace applications (drop-in vapor degreasing), the use of HSP can assist in combining less than ideal cleaners together to create a more viable cleaning solvent mixture. More details on the overall workings of the HSP values can be achieved by using the Hansen Solubility Parameters in Practice software available from Hansen-Solubility.com

Together all of these additional resources discussed can help to fill the chambers of the gun in the search for the TCE Silver Bullet.

Chapter 10

If No Silver Bullet Where To Run To, Who Else To Turn To

Abstract

The TURI Lab is not the only game in town when it comes to providing technical assistance for cleaning projects. The previous chapters have given a lot of information on how TCE could be reduced or eliminated. Throughout the United State, there are regional pollution prevention groups and individuals that can help make the process go smoother. Relying on internal expertise may work for some, but many times external resources will be needed, especially for smaller companies.

The previous chapters have given a lot of information about how TCE could be reduced or eliminated and the third party agencies that can also help pick a viable alternative to TCE for cleaning needs. There are also other places to turn in many cases if a company is able to reach out and see what exists in their area.

Technical assistance can come in many forms from universities, trade magazines, trade show education sessions, local trade groups, consultants, non government organizations, pollution prevention programs, state agencies, and federal resources. Also, in many cases you may have an internal resource or champion that works for you currently, and that is where this chapter starts.

The following resources are not intended to be an exhaustive listing, but more of an illustration of the types of resources available in the pursuit of the TCE Silver Bullet.

Internal Resources and Champions

Many pollution prevention projects, like eliminating TCE in cleaning, are collaborative processes and require buy-in from employees in order for the project to move forward effectively. Many times companies do not even know that they have a diamond in the rough internally, which is why companies need to communicate about a project to the whole company.

By sending out a companywide email to update all workers on Pollution Prevention (P2) projects, the company is letting them know about the work environment and that the company is moving forward with a project. This may elicit interest and setup less opposition to change when everyone seems to be a part of the project.

This may also provide a good opportunity to ask employees if anyone has run a "green team" or done a similar project in the past at this or another company. This is a resource often over looked, but one that can be invaluable, so make sure to reach out and ask.

Other company resources are the EH&S department. They will most likely jump at the chance to work on a project rather than spend all their time on reports and regulation paperwork. In addition to the EH&S department, companies may have a sustainability group. They may look at many areas related to greening or helping your company become more environmentally conscious. Many people on these volunteer groups are interested and passionate about pollution prevention and protecting people and the environment, and they may hold the expertise the company seeks for the cleaning project at hand.

Local Universities

Many P2centers are located at or near, or collaborate with universities in order to use the great resources of researching professors and students. This is an excellent place to look for help with a P2 project if you have no idea where to start. Universities are a great place to find motivated collaborators that will partner at little or no cost to you in order to do the work. They are also good places to find interns and summer workers that could end up being great full-time employees when they graduate, so do not overlook universities as resources.

Business Specific Magazines and Newsletters

These can be excellent resources for technical advice and articles on companies doing P2 projects just like you. In many instances you can email or contact the company in the article or the technical expert writing the column for more information. In addition, the publisher and or editor are generally well-versed in the industry they write about, sponsor and attend shows and conferences, and know the major players in the industry, and they may be able to help you with leads to help with your P2 project.

One media group that has many publications in the metal working industry is Gardner Media, http://www.gardnerweb.com. They publish the well known Product Finishing Magazine as well as Modern Machine Shop, Production Machining, Mold Making Technology, Gear Production, and Automotive Design and Production among others.

Many magazines also publish newsletters that they deliver electronically on specific topics in an industry. Check out their websites for more information.

Vendors

Yes vendors want to sell you things, but many good ones have evolved into technical assistance partners also. If you have a good vendor you work with now but need help switching a process or chemical input in your manufacturing process, they are a good place to start. With good vendor collaboration, they ideally will want to keep your business and help you with the changes you want to make. Remember, your money can talk and you can walk if you are not getting the services you need. Many companies sell the same products, and now is the era of added value. Make sure you get that added value; you are responsible for getting it.

Many salesmen have often been in the industry you are in for many years and have had different positions along the way, and they can speak with experience about many aspects of the sector for you. Their expertise is often wide and some of the best you will find. Companies of a decent size also have their own laboratories and chemists who can work with you and the salesman on your issues or challenges. If you are a large enough, customer products can often be tailored to your needs.

Many sales people and employees of the vendors you work with are members of local trade associations and have a pool of experts to pull from as well. Time is a resource for sure, but attending some of these local groups can help you connect further with your vendors and their employees to find solutions. Networking is never a waste of time.

National and Local Trade organizations

Trade associations can provide a wealth of information. They often put out newsletters or magazines. Check their websites for information and sign up for them. They also may have a national or regional conference that may also include a tradeshow with vendors and educational sessions on topics important to your P2 project.

Two examples of these organizations that have resources associated with TCE and solvent cleaning are the Society of Manufacturing Engineering, known today as SME, and the National Association for Surface Finishing (NESF).

SME, http://www.sme.org, works on all facets of manufacturing. It holds education sessions, training and different events throughout the year around the country. For example, in the Northeast they put on EASTEC in Western Massachusetts in May of every other year. Check out their site for more information.

NESF, http://www.nasf.org, is the national association that holds training, information sessions and a national trade show called Sur Fin every Spring/Summer in June in the Midwest. They can provide technical information, policy and regulation news, and other resources.

Along with the national meetings and conferences, many trade associations including SME and NESF, have local chapters that are active in education and networking events throughout the year. They are often inexpensive to join and offer great connections and opportunities to find expertise near you.

Consultants

Yes it is true; consultants can cost a lot of money but may be worth it. Environmental consulting firms can be a one-stop shop for all your needs.

They have experts in many fields who have extensive education and practical work experience in the areas with which your company may need help. Their websites can be a good place to start, and if interested, calling or emailing for a consultation never hurts and is often free.

Lobbyists

They are in the fight every day for different industry sectors and have their finger on the pulse of policy and regulations that could affect your company. Often they are connected with the trade associations.

P2 Programs and Centers

Throughout the United States, there are regional pollution prevention centers that are funded in part by grants from EPA. The mission of these groups is to advance pollution prevention as a cornerstone of sustainability. They build networks, deliver P2 information, and measure P2 program results. The effectiveness of the network is tied to the expertise and diversity among the regional centers and wide range of the intended audiences served. These include states, local governments, technical assistance providers, businesses, educators, and nonprofit organizations. The groups and the states they serve are listed in Table 10.1. A summary of each of the programs is provided.

Northeast Waste Management Officials' Association (NEWMOA)

NEWMOA is a non-profit, non-partisan interstate association whose membership is composed of the state environmental agency directors of the hazardous waste, solid waste, waste site cleanup, emergency response, pollution prevention, and underground storage tank programs in Connecticut, Maine, Massachusetts, New Hampshire, New Jersey, New York, Rhode Island, and Vermont. NEWMOA's mission is to develop, lead, and sustain an effective partnership of states that helps achieve a clean, healthy, and sustainable environment by exploring, developing, promoting, and implementing environmentally sound solutions [35].

Environmental Sustainability Resource Center (ESRC)

ESRC provides pollution prevention technical assistance and serves Regions 3 and 4 with comprehensive online resources, news, and information about pollution prevention. The center utilizes regional and national expertise to provide P2 information that is critical to state environmental agencies, businesses, educators, technical assistance providers, and the general public. ESRC believes that helping organizations move toward environmental sustainability will reduce costs, minimize ecological impact, and help conserve, protect, and maintain the environment [36].

Great Lakes Regional Pollution Prevention Roundtable (GLRPPR)

The GLRPPR is a professional organization dedicated to promoting information exchange and providing networking opportunities. GLRPPR's membership consists of assistance providers, business, industry, governmental agencies, NGO's, nonprofits, consultants, vendors, universities, etc. from Wisconsin, Minnesota, Illinois, Indiana, Ohio, Michigan, Pennsylvania, New York, and Ontario. GLRPPR also has members that are located outside of the Great Lakes region, but have similar environmental priorities and concerns [37].

Zero Waste Network

The Zero Waste Network is a group of environmental professionals dedicated to finding money-saving options for conserving our natural resources. It is a collaborative project of the U.S. EPA, the University of Texas, and regional environmental agencies that works to identify pollution prevention options for large and small businesses as well as consumers. The Network hosts databases with over 500 success stories, a material exchange service, and over 1000 BMP's supplied by industry and partners. The Network trains nearly 300 environmental professionals from industry, local government, and other entities to prevent pollution at its source with EMS's, Lean and Green Processing and process mapping [38].

Pollution Prevention Regional Information Center (P2RIC)

P2RIC strives to improve resource sharing between the programs, businesses, and agencies that provide waste reduction services and expertise to business and industry. Housed in the first LEED certified building in the University of Nebraska system at the College of Business Administration, P2RIC emphasizes partnerships to promote sound business strategies for a sustainable future [39].

Peaks to Prairies Pollution Prevention Information Center

In cooperation with EPA Region 8 states, Peaks to Prairies encourages adoption of pollution prevention practices by citizens, small businesses, and local governments. The Center offers access to current information and contacts, encourages collaboration and leveraging of resources between programs, and builds information systems to enhance information dissemination [40].

Western Sustainability and Pollution Prevention Network (WSPPN)

WSPPN is a strategic alliance involving local, state, federal, and tribal pollution prevention programs throughout EPA Region 9, which includes the states of California, Nevada, Arizona, Hawaii, the Tribal Lands, and the Trust Territories of Guam and American Samoa. WSPPN has documented service delivery to over 175 programs through its annual conference and to 580 organizations through its webinar service. WSPPN is administered by the Business Environmental Program at the University of Nevada, Reno College of Business affiliated with the Nevada Small Business Development Center (NSBDC) and is uniquely positioned to lead WSPPN as a statewide technical assistance provider of advisory services for environmental compliance, pollution prevention, sustainability, energy efficiency, renewable energy, and clean technology development [41].

Pacific Northwest Pollution Prevention Resource Center (PPRC)

PPRC is the leading regional nonprofit that provides practical, on-the-ground technical assistance to businesses, public agencies, and nonprofits that are seeking to conserve resources and improve economic performance. PPRC believes that environmental and economic vitality go hand in hand, and that both are necessary to protect the high quality of life enjoyed in the region. PPRC's focuses are helping to reduce toxic pollutants in the environment and improve energy efficiency in regional businesses and manufacturers [42].

Table 10.1. Pollution Prevention Resources by EPA Regions

Regions	Name, Location	States
1 & 2	Northeast Waste Management Officials' Association 129 Portland St., Suite 602 Boston, MA 02114-2014 617-367-8558 #306 617-367-0449 (FAX) Contact: abray@newmoa.org http://www.newmoa.org/	CT, ME, MA, NH, NJ, NY, RI, VT
3 & 4	Environmental Sustainability Resource Center Kentucky Pollution Prevention Center, University of Louisville Louisville, Kentucky 40292 502-852-0965 502-852-0964 (FAX) Contact: info@esrconline.org http://esrconline.org/	AL, DC, DE, FL, GA, KY, MD, MS, NC, PA, SC, TN, VA, WV
5	Great Lakes Regional Pollution Prevention Roundtable Illinois Sustainable Technology Center One E. Hazelwood Dr. Champaign, IL 61820 217-333-8946 217-333-8944 (FAX) Contact: glrppr@istc.illinois.edu http://www.glrppr.org/	IL, IN, MI, MN, NY, OH, PA, WI, and Ontario, Canada
6	Zero Waste Network University of Texas Arlington Center for Environmental Excellence 9390 Research Blvd, Suite 300 Austin, Texas 78759 512-904-2281 512-904-2288 (FAX) Contact: admin@zerowastenetwork.org http://www.zerowastenetwork.org/	AR, LA, NM, OK, TX

Regions	Name, Location	States
7	Pollution Prevention Regional Information Center University of Nebraska at Omaha 6001 Dodge St., RH 308 Omaha, NE 68182 402-554-6257 402-554-6260 (FAX) Contact: contact@p2ric.org http://p2ric.org/	IA, KS, MO, NE
8	Peaks to Prairies Pollution Prevention Information Center Montana State University Extension Service PO Box 173580 Bozeman, MT 59717 406-994-3451 406-994-5417 (FAX) Contact: information@peakstoprairies.org http://peakstoprairies.org/	CO, MT, ND, SD, UT, WY
9	Western Sustainability and Pollution Prevention Network 6100 Neil Rd., Suite 400 Reno, NV 89511 775-689-6678 775-689-6689 (FAX) Contact: dwalden@unr.edu http://wsppn.org/	AZ, CA, HI, NV
10	Pacific Northwest Pollution Prevention Resource Center 1402 Third Ave. Suite 1420 Seattle, WA 98101 206-352-2050 206-352-2049 (FAX) Contact: office@pprc.org http://www.pprc.org/	WA, ID, OR, AK

University Resources

The University of Tennessee Center for Clean Products (CCP)

In 1992, in anticipation of the emerging national and global trend toward product stewardship, pollution prevention, and green consumerism, a group of forward-thinking researchers at the University of Tennessee (UT) established the Center for Clean Products (CCP). CCP is a multi-disciplinary research center dedicated to the development, evaluation, and adoption of clean products and materials. The Center is housed within UT's Institute for a Secure and Sustainable Environment. CCP staff members bring a combined 40 years of experience to partnerships with industry and other stakeholders, working to improve the environmental performance of a wide range of consumer products and manufacturing systems. Through these partnerships, CCP has assisted industry leaders in adopting a life-cycle perspective, incorporating environmental concerns into the earliest phases of product design, and making clean production a cornerstone of their corporate culture [43].

The New York State Pollution Prevention Institute (NYSP2I) Rochester, NY

Headquartered at Rochester Institute of Technology (RIT), NYSP2I is a partnership between RIT, Clarkson University, Rensselaer Polytechnic Institute, University at Buffalo, and the 10 NYS Regional Technology Development Centers. The mission of NYSP2I is to make New York State more sustainable for workers, the public, the environment, and the economy through: reductions in toxic chemical use; the efficient use of raw materials, energy, and water; and reductions in emissions to the environment and waste generation. NYSP2I is an integrated program of several major elements: technology research, development, and diffusion; outreach; professional training; a community grants program; and academic educational programs [44].

University of Illinois at Urbana-Champaign, Illinois

Energy and Sustainable Business Programs performs applied research associated with technologies that improve operational sustainability with respect to materials, energy, and water usage; implementation assistance associated with innovative practices and technologies; demonstrations of more sustainable practices both in the field and in the lab in order to improve adoption rates; and development of three patented technologies for industrial fluid purification [45].

Lowell Center for Sustainable Production, UMASS Lowell, Lowell, MA

The Lowell Center has set its long-term goal to redefine environmentalism and occupational health and safety while also demonstrating how these concepts are compatible with new systems of production and consumption that are healthy for workers, environmentally sound, economically viable, and socially accountable. Through its solutions-oriented and collaborative research projects, the Lowell Center has become an internationally recognized resource for developing and piloting the concepts of sustainable production and consumption [46].

The Toxics Use Reduction Institute, (TURI), UMASS Lowell, Lowell MA

The Toxics Use Reduction Institute (TURI) at the University of Massachusetts Lowell was established by the Massachusetts Toxics Use Reduction act, (TURA) of 1989. Its unique position of being located at a state university gives the institute a wide base of collaboration with students and university professors and researchers. TURI provides resources and tools to help make the Commonwealth a safer and more sustainable place to live and work. TURI collaborates with businesses, community organizations, and government agencies to reduce the use of toxic chemicals, protect public health and the environment, and increase the competitiveness of Massachusetts businesses. Their laboratory is unique in that is does testing on less toxic alternatives and equipment as well as R&D, outreach, training, and field and technical assistance around implementing less toxic alterative chemicals and processes.

Institute for Research and Technical Assistance, IRTA

IRTA is a nonprofit organization located in Los Angeles, CA, with extensive expertise in research and technical support and guidance into safer alternative chemicals and processes. They have been doing work in many areas for years and have written many technical reports that help many sectors in their pursuit for less toxic alternatives. Some of this important work has occurred in the sectors of dry cleaning, electronics, batch load cold cleaning, vapor degreasing, hand wipe cleaning, precision cleaning, and more. Their goals are threefold; to identify, test, and demonstrate safer alternatives with an emphasis on solvents, to demonstrate and evaluate new and emerging technology and processes in the field and with the end users, and to work with regulators on policies.

FEDERAL RESOURCES

EPA's Pollution Prevention (P2) Program

The broad mission of EPA's Pollution Prevention (P2) Program is to prevent pollution at the source, promote the use of greener substances, and conserve natural resources, which are critical steps towards achieving a sustainable society. P2 is a cornerstone of the sustainability movement, which strives to ensure that our children and grandchildren inherit a world that is as good as it is today, or preferably better.

P2's work is authorized and funded by statute; multimedia by definition; well leveraged by state and local partners and networks; and capitalizes on established trust relationships with U.S. businesses that appreciate that environmental protection and prosperity are not conflicting goals. The program's authority comes primarily from the 1990 Pollution Prevention Act (PPA) [47].

The National Pollution Prevention Roundtable

The National Pollution Prevention Roundtable is the largest membership organization in the United States devoted solely to pollution prevention (P2). The mission of the Roundtable is to provide a national forum for promoting the development, implementation, and evaluation of efforts to avoid, eliminate, or reduce pollution at the source [48].

U.S. Small Business Administration (SBA)

U.S. Small Business Administration (SBA) was created in 1953 as an independent agency of the federal government to aid, counsel, assist, and protect the interests of small business concerns, to preserve free competitive enterprise, and to maintain and strengthen the overall economy of our nation. SBA recognizes that small business is critical to economic recovery and strength, to building America's future, and to helping the United States compete in today's global marketplace. Although SBA has grown and evolved in the years since it was established in 1953, the bottom line mission remains the same. The SBA helps Americans start, build, and grow businesses. Through an extensive network of field offices and partnerships with public and private organizations, SBA delivers its services to people throughout the United States, Puerto Rico, the U. S. Virgin Islands, and Guam [49].

More Resources Available

Many of the programs listed in this chapter could be under more than one heading. We also recognize that this was by no means an exhaustive list of resources. It was an effort to highlight some of the organizations across the country. Hopefully the listing is a place to begin for those struggling to get started on the path to reducing their use of TCE.

Chapter 11

OBSTACLES, BARRIERS AND ISSUES, OH MY!

ABSTRACT

Changing a process or a chemical that is integral to the manufacturing of something there are often issues, barrier, challenges and obstacles to overcome or deal with. These can generally be overcome with a good team working on the project as well as taking the appropriate amount of time needed to meet all internal department questions and concerns as well as client and or customer needs.

LET'S GET THE PARTY STARTED

When talking about changing a process or a chemical that is integral to the manufacturing of a product there are often issues, barriers, challenges, and obstacles to overcome. These can be actual physical, philosophical, management, training, client, and technical issues or challenges. However, these challenges can generally be overcome with a good team working on the project as well as by taking the appropriate amount of time needed to meet all internal department questions and concerns as well as client and or customer needs. Sometimes the process goes fast and smooth, but more often than not, a successful change can take a while. When done right, the changeover will benefit the company, workers, and the community.

Areas this chapter will look at under the heading of barriers, challenges, and obstacles to change will be the ones seen the most in the technical assistance arena and those from the facility managers or head engineers who are usually the leads on these types of projects. These more common areas of concern, or where a project can hang up, include: full management buy-in at all levels; cost and capital investment by the company; putting together a good team; long term commitment to a project; deciding how far to take the project; technical issues and training concerns; space; regulations current and future; customer specifications and testing needs for the switch approval; solvent compatibility issues; and new process compatibility issues. This is not an exhaustive list of challenges by any means, but the list contains some of the most common issues that arise. Every company, process, and situation is unique and will have its own set of concerns to address. Let's take them one-by-one and discuss.

Full Management Buy-in for the Project

This should be self-explanatory, but it is not always obvious. Sometimes an issue can occur such as a regulation change, a worker injury, a workman's comp filing, a piece of machinery needs replacing, retiring or retrofitting, an OSHA or some other inspection, or maybe the company wants to have a better EH&S profile and advertise it. These can all be the reasons why a project to change the use of TCE (and other hazardous solvents) can begin to be discussed. Depending on the reason for the project there can be different levels of management buy-in. Generally a project can only move so far without upper management being on board. This is because they usually control the purse strings and also allow the time for the project to be carved out of workers' regular schedules. If management is not on board, then starting a project to replace TCE can be next to impossible to move forward.

Once the issue or reason for the project is identified, communication and building a case for the project is key. Say the EH&S manager is aware of regulations that are coming or changes in reporting that will impact the company, such as time to do the reporting, cost to meet the regulations, and possibly more training and cost to the company, the case could be made to look at a different process to get out of TCE on more than one level. For example, in Massachusetts the Toxics Use Reduction Program designated TCE as a High Hazard Substance beginning in reporting year 2008 [50].

What this did was change the reporting requirements for companies using TCE. If a company had 10 FTEs (full-time equivalent employees) and used 1000lbs (about a drum and a half of TCE), then they would be brought into the TURA system and need to do paperwork on their use, a plan on how they could reduce or eliminate their TCE use, and pay fees to the state. In this instance, switching away from TCE would be beneficial to the company in that they would not have to participate in the above mentioned tasks. Now deciding if there is money in the company budget to start the project is the next challenge.

Cost and Capital Investment Commitment

Once a project is seen as important by the company with some, most, or all of management being behind the idea of the new project, the question has to be: is there money to do the project in full, or will this be an exercise in vetting the project's costs and other details first? Either is a good start, but identifying where everyone is and what the expectations will be are both crucial and take time. Since investigation of a possible project can cost money in the form of employee time to investigate, the task shouldn't be taken lightly. It is important to talk to people, and do research. Everyone knows that time is money.

Switching to a drop-in solvent for existing machinery costs money and time as does changing the whole process. The costs may be very different but funds still need to be freed up in the budget to meet the related expenses. This is where upper management's involvement and understanding of the project being done is crucial. If possible, a member of upper management should be asked to be on the project team and kept apprised of the project. This is where assembling a team comes into play.

Picking members of your Project Team

When working on a project to switch from TCE to another less toxic solvent or process, a good team should be put together from as many departments in the company as possible in order to consider all points of view on the changes. This team can help head off potential issues in the project by working together and giving insight from each member's particular department. Members of the team could include people from the management, sales, purchasing, manufacturing, engineering, assembly, and EH&S departments. In addition, current and potential vendors and any local or state technical assistance providers could be added to the team if such resources exist. Sometimes local college professors and students can be of help, as they are always looking for projects for research. Local trade associations sometimes have expertise that can provide assistance as well.

Once the team members have been selected, weekly, monthly, or other regularly scheduled meetings should be established in order to help alleviate scheduling issues. Everyone is busy, so sharing information electronically can also help. Meetings may be done as working lunch meetings to make them easier for people to attend or coupling them with mandatory company meetings such as staff meetings when more people are likely to be at work.

This process should be set up early on and designed to be as easy and unobtrusive as possible to keep the team excited and committed to the meetings. If certain members cannot commit the time, then it may be beneficial to pick other members from the same department to back them up or take over and be the primary department member so attendance can be maintained. Barriers or challenges here can be maintaining a meeting schedule and keeping members of the team on task. Being firm, but accommodating is necessary as the job must go on. The facilitator or leader of the project must keep everyone on task, stick to a timeline if possible, and help get everyone on board to define the project and the steps necessary to complete the project.

DEFINING A PROJECT AND HOW FAR TO GO WITH IT

Once you have management buy-in, possible capital encumbrance for the project and a team in place, the project now needs to be discussed and the scope decided on. Again this can seem self-explanatory, but with different people on the team, from different departments, with different perspectives, this task can sometimes be the biggest challenge and may take a fair amount of time. Some of this can be done by email, but very often face-to-face discussion can diffuse tensions that may arise between departments trying to decide what is most important for the project and how it will be done. Other decisions to be discussed may be who will research what, who will be responsible for what deliverables, and other subjects of this nature.

Remember the project is company specific, and the internal culture may dictate different paths. While others outside the company can help, the project must rely upon internal expertise of the team. There is no right or wrong way to do this, just the way that works best for each company. Patience is a virtue here, and the old practice of counting to ten can go a long way to keeping the project alive. It is also important to try to keep things light by being complimentary and encouraging the sharing of ideas. Creating a safe place to share ideas will keep the team loose and make them more inclined to work together and help. Remember that when all else fails, you can bribe the team with food, especially donuts.

Solvent replacement projects often mainly fall into two areas: drop-in replacements using the same process and equipment the company already has in place; or a full process change to new equipment and chemistry. This decision on the project scope or type generally dictates the rest of the project and is not to be taken lightly. A drop-in replacement for a solvent may be a shorter-term fix for the issues that brought the project to light in the first place and may solve many of the company's major issues in using TCE in their process.

Drop-in replacement projects for solvent replacement mainly happen for a few reasons. The first and most common reason occurs when a company has invested significant capital into equipment, such as buying a new high-end degreaser or retrofitting an existing vapor degreaser. Here it is understandable that a company would want to use the capital they have already invested. Other major reasons to opt for a drop-in solvent replacement project are overall capital investment needed for a drop-in project versus a full process change and fear of what a change in solvent may mean to the business (rejects, training, client happiness, acceptance etc., client specifications such as military, aerospace, and FDA for parts that will go into medical devices). In these cases a drop-in replacement for the solvent is much easier for many people to handle.

Full process changes for solvent switches can be done with careful planning, time, and capital. They are more intensive for sure, but for the progressive companies that are ready for the challenge making this switch can be exciting and worth the final outcome. With a good team these challenges can be worked through one-by-one. Many of the obstacles include: management and worker buy-in for the change; client acceptance of the change, especially if it includes going to aqueous solutions and introducing water to the process; training; space for a new system and its components such as rinsing and drying operations; piloting and testing the process enough to prove the new equipment and chemistry works in a lab and in real time; time and commitment to a long project for all team members; money or capital;

dealing with changes or failures during the process that may arise; fear of the unknown; and wanting to do things the same way as always.

When choosing to go all the way or from 0 to 60 in one step, one of the main drivers needed is available liquid capital. In addition, EHS is a main topic of discussion and concern for the whole company, including wanting to increase good public perception of a company or be a better community member; receiving a pollution prevention grant or other capital source that includes technical assistance; being in a state that offers technical assistance and resources to help; regulations and costs of both state and federal programs to use the toxic solvent are becoming cost prohibitive; and whether your supply chain is involved in driving the switch.

Sometimes the process of switching from an effective, albeit hazardous, solvent can be iterative and that is ok and valid. The step wise process is often done in order to do a project such as this in bite-size, digestible pieces that all participants and departments and workers affected can handle. The segmenting process allows for more opportunity for feedback as each step is completed. Remember there is no wrong way to do a project as long as it works for your company and keeps your workers, community, and environment safe. Taking your time, doing it right, and at a pace everyone can handle, is important to the project's success no matter what avenue you are choosing or investigating.

TECHNICAL ISSUES AND TRAINING

When considering a project that will switch an integral part of any manufacturer's or shop's process there will always be technical and worker training issues to address and research. There will be pros and cons to whatever option a company chooses, and it may be a good idea to have your team brainstorm ideas for the change. This is where a diverse team from various departments of the company will come in handy. When doing this brainstorming, set guidelines for participation. As previously mentioned, the most important guidelines to set are that no ideas are bad ideas and that the group is working in a risk and criticism free zone. There will be time after this to talk about the options and ideas and eliminate those way out of scope or possibility due to technical, time, cost concerns, or other obstacles.

To help demonstrate, Table 11.1 focuses on drop-in replacement versus total process change projects. These are generally the main types of projects for solvent replacement. The table highlights the main pros and cons for each type of solvent replacement project as seen from the TCE replacement perspective.

As you can see from Table 11.1, there are compelling reasons to change and not to change, go all the way or go part way. Again, what is right for one company may not be right for another, but identifying your technical and customer needs as well as the drivers for your project and to change or not change will help guide you through the process. A project can also be part of each of these options and done in steps. It is never bad to at least talk about it, research it, and prepare for a change now or in the future.

Customer Cleaning Specifications and Needs

The purpose of being in business is to make money. Yes there are companies that want to do good work, make superior products, be a good community member, and many other reasons for being in business, but overall the reason is making money. Keeping clients and making a good cost effective product of some kind for the industry you serve is the main goal. All other reasons for having a business cannot exist if the company closes. So any project, no matter how small, has to keep staying in business and keep its clients in mind.

With this idea in mind, many companies that use TCE or other hazardous and regulated solvents in their process to clean may service clients that are in the military, aerospace, and medical devices sectors. These sectors are heavily regulated and have many specifications that need to be followed by their supply chain in order to keep their business. For these reasons, when switching from TCE to an alternative process or solvent these particular requirements must be addressed in addition to working with clients, suppliers, and the company's supply chain as a whole. Furthermore, knowing the needs of your customers and controlling rejects is crucial to keeping these customers. Often times if these types of clients are your company's major customers, then what they want may guide and make many of your decisions for you. The customers may actually drive decisions about what solvent or process you can use and what amount of testing needs to be done to prove to a client that the new solvent or process will meet their specifications.

Table 11.1. Solvent Replacement Projects Comparison

Process Change	Pros	Cons
Vapor Degreasing to Vapor Degreasing (Drop-in)	Technically familiar Less EHS training on new chemical needed Vendor can often help with change Major cost change attributed to new chemical mostly No major client concerns or testing generally needed (check specs) Generally alternatives available by industry, substrate, soil, or all Use current equipment Use current space or footprint in facility Easier for management and workers to swallow Can get company out of regulations at least temporarily Keep same hazardous disposal system	- Not a major EH&S improvement - Cost of new solvent - May still have some regulatory costs for state and federal government - Not a switch that can be heavily publicized - Still may have to do significant testing for client specs -Will take time to research and test alternatives to make sure they work on all substrates and soils - New regulations may pop up down the road for replacement solvent as in the case for TCE to nPB substitutions. Now nPB is regulated by many states and the federal government after being touted as a replacement. - Disposing of old solvent in drums, buckets and stills
Vapor Degreasing vs Total Process Change, Aqueous	- Major EH&S Change, much safer for workers and environment - Good PR opportunity for company with community and workers - Increase worker moral - Less PPE needed - Less regulatory paperwork and costs, saving labor costs and time and money overall	-More technical concerns - Introducing water to the system - Need to add a rinse and a dry most likely - Worry over flash rust and oxidation of parts possible -Space issues, may need more for new system, this has potential costs - Capital needed exponentially more than for just solvent replacement, need to buy equipment, costs to put the system in, time to research and see other systems in process, test, pilot, and train - Customer testing will be required, may not meet their specs or take significant testing and time to meet them. -Water disposal, permits, or zero discharge concerns -Time for project
Vapor Degreasing	- Change processes that are not	- Some water introduced to process

Process Change	Pros	Cons
(VD) vs. Partial Process Change, VD and New Process	integrated with complicated client specs easily - Increase some EH&S specs significantly - Keep part of process familiar to everyone - Less overall testing, training, and tech issues than with full switch - Part of project can be done faster than the other, so possible to get out of some regulations easily - Some good PR - Possible good option for many - May be able to downsize VD equipment, sell current one to raise capital to buy smaller one, and maybe help pay for new equipment	- Some testing and piloting will be needed - Still significant amount of capital and time to do project, in between drop-in and total process change - Not going all the way, but further along than drop-in replacement - Still a lot of training and research needed
No Changes, Keep Current Process	- No changes to deal with workers, no new training - No cost changes or capital investment needed for process - Same PPE	- Same exposure and EH&S issues - Will have to pay and comply to state and federal regulations and permitting - Possible worker injury, illness, and works comp claims - Maintenance and upkeep of machinery

Complying with client needs, making sure they are happy and satisfied while keeping their business, often dictates how long a project may be. It also affects how the testing and process validating for the project proceeds. Many times this is where the baby steps approach is taken to prove that a new solvent or process can work. Often trust can be earned and will make a long-term project easier with this type of plan.

Solvent Compatibility

This is always a concern when switching from TCE to any other drop-in replacement. Companies use TCE, PCE, nPB, methylene chloride, and other similar hazardous solvents because they work, have great solvency, are more forgiving to use, are cost effective, and are the devil they know. From a technical provider aspect it can be understood why a company would keep these solvents, but from the EH&S, pollution prevention specialist aspect, companies need to be urged to at least look at safer alternatives.

Once we leave the cast of characters mentioned above, the drop-in replacement solvents tend to increase in cost and decrease in solvency. This is where the testing and research comes in to identify a solvent to put in your current degreaser that will work for your company's substrates and soils. This can be harder than thought when leaving the realm of "super" solvents. Time and attention must be given to all combinations of soils and substrates needed to be cleaned as well as the new technical aspects for the degreaser such as temperature and time to clean in order to make a drop-in solvent replacement project successful. Working closely with a vendor may be needed.

PROCESS COMPATIBILITY - ADDING WATER TO THE MIX

When moving to a whole new process, assuming space, capital, and commitment needs are met, then technical needs come into play. Here, generally going to an aqueous or semi-aqueous system can introduce water and new processes like rinsing and drying to the system. These are the challenges or barriers. Many times engineers and others will say emphatically, that water cannot be used. This can be where the challenge begins. Is this true? Can it be verified? Is it just going to be a long hard road and process or can water really not be used? This is where research and talking to clients and the supply chain a company is part of is necessary. Trade associations and local meetings with other companies may also work and help. Talking peer-to-peer may help to find out if someone else was able to work around a challenging specification or issue. Networking can help solve many problems.

Many times, cleaning parts is not the issue. The cleaning can often be accomplished with very mild cleaners; it is the procedures that take place after the cleaning that dictate the process. This is where water introduction can become an issue. Vapor degreasing solvents will clean a wide range of soils and substrates. The process is also one that does not generally need a rinse or a dry of any kind, saving time and steps. In some cases, the solvent vapor degreasing also leaves behind a layer that will protect the part from flash rusting or oxidation. This enables the part to be stored, sit, or be protected during transport to the client.

When you introduce a whole new process that will include different mechanical cleaning processes such as spray wash, soak, and ultrasonics or a combination of all of these, water is a big factor in the equation. With parts with small blind holes or complicated geometry such as on gears, water can be a scary thing to use. The reason for this is that water has a much higher surface tension then solvents and may be caught in small blind holes. Water also may not be able to remove any or all of the metal fines in the holes formed during manufacturing. The water that is trapped or left behind may cause rusting deep inside a part or on the surface and cause major issues of rejects of batches of parts and possible loss of business and clients. There goes the money.

This issue can be overcome usually with the addition of products such as wetting agents. Further additions to the process such as high-filtration, drying equipment will help control the water issue. In addition, setting up a highly automated system with good operators who understand it to run it minimizes the impact. Many times an aqueous system will work if time, money, and patience are given to the project. It will not hurt if great technical experts are part of your team.

FEAR

Fear is the universal barrier and challenge. It is understandable and very real for many. Will this work? Will it be a waste of money? Is it worth the effort or should I just pay to use TCE? Does the company have the expertise to do this? What if a mistake is made? Will the clients go for it? What about workers, engineers, or everyone and anyone? With a good team in place, planning, resources, and time, the fear can be mitigated and managed. If all else fails, a good margarita may help.

SUMMARY

Even though the chapter ended with fear, it usually is a common thread to all other obstacles. However, by getting all parties involved and committed to work together to uncover what these fears are and how to address and overcome them will result in a successful project. Getting the resources squared away and all of the technical specifications outlined will help upper management buy-in to the change and provide the needed capital investments.

Chapter 12

CASE STUDIES OUTLINING THE STEPS TAKEN TO REPLACE TCE

ABSTRACT

Case studies provide an illustration of how the process of solvent substitution takes place at the TURI Cleaning Lab. An effort has been made to provide examples of switching from TCE to an aqueous cleaner, to a semi-aqueous process, solvent replacement with new process and a straight drop-in substitution. Not every laboratory tested product will end up becoming a successful substitute.

If a product is green but does not work, then so what; nobody is going to use it. The long running fear of green products has been tied to performance. Early on, this was a justifiable concern as many of the initial green products did not work as well under the same conditions as the traditional products. More often, extended cleaning times or additional forms of agitation were needed to ensure proper cleaning levels were achievable. In an effort to address the shortcomings, independent performance testing was introduced to demonstrate that the next generation of green cleaners was as effective as conventional products and more effective than initial green products.

The following case studies provide an illustration of how the process of solvent substitution takes place at the TURI Cleaning Lab. An effort has been made to provide examples of switching from TCE to an aqueous cleaner, to a semi-aqueous process, solvent replacement with new process, and a straight drop-in substitution.

Not every laboratory tested product will end up becoming a successful substitute.

This was what happened in the tube manufacturing project. The TURI Cleaning Lab identified several possible alternatives that were not selected. However, the company was exposed to the possibility of not needing to stick with the old faithful. Building on the lab's research, they identified and piloted a drop-in solvent on their own.

The capacitor cleaning case study looks at a company that needed to use their recently purchased equipment and had to focus on a drop-in solvent. After 10 years of use and the changing regulatory status of their initial drop-in solvent, the company returned to the TURI Cleaning Lab to help them switch to an aqueous cleaning system. This stepwise process has been the model for others not quite ready to go away from solvents completely.

The third and final project looked at TCE replacement for a company working in the military/aerospace sector. To substitute the use of TCE, they needed to document performance of the alternatives internally and then work toward gaining the acceptance of their clients.

12.1. Case Study Tube Manufacturing - A Light at the End of the Tube for TCE to Semi-Aqueous

Summary

From its earliest days, the Tube Company has been an important supplier of small diameter, extremely close tolerance metal tubing to industries in the aerospace, nuclear, oil and gas, power generation, and medical equipment industries. The product line being investigated included materials being used in implantable medical devices and instruments. The company has a commitment to social responsibility and to eliminate the effects the company's work has on the environment. Health and safety issues are taken extremely seriously by both management and employees.

Background and Problem Statement

The TURI Cleaning Lab was contacted by the Tube Company to help determine the best method of cleaning inside diameters of narrow metal tubes.

Cleaning was being conducted at the facility to avoid metallurgical contamination in their annealing process. Their motivating factor was that their current process was labor intensive and more importantly, it emitted TCE. The implantable medical devices and instruments that they make were made from mostly stainless steel but could include various nickel alloys as well. The tubes had inside diameters that ranged from 0.005" to 0.08" as shown in Figure 12.1 Cleaning took place in batches of 300 and around 6000 parts were cleaned per week. The process used just under nine tons of TCE per year.

Figure 12.1. Tubes to be Cleaned.

Table 12.1. Soils Used for Tube Manufacturing

Name	OD	ID	Soil ID	Remarks
Rirchards Apex #226	Y	N	1	Applied to tube OD during drawing
Rirchards Apex #201	Y	Y	2	Applied to tube OD during drawing. Also, for heavy draws, mandrel rod tips are dipped in 50:50 mix of #201:Lard oil prior to loading into tube
Lard Oil	N	Y	3	Mandrel rod tips are dipped in lard oil prior to loading into tube. Often mixed with RA #201
#5 nPB Supercoat	Y	N	4	
3:1 nPB Supercoat	Y	N	4A	
Premco #6	Y	N	5	Lubricant used to flood swager dies during tube expansion process

The manufacturing process used several soil types. The soils and uses are listed in Table 12.1. The soils listed were then prioritized by the client to help direct the alternative selection process and performance testing.

Phase I Product Selection

Typically, The TURI Cleaning Lab initially attempts to identify aqueous-based products for most cleaning applications. In the event that no aqueous products are found to be effective, semi-aqueous products are then selected for evaluation. Ten products were selected from the lab's database, based on client supplied process information, past product performance, and safety screening values. Table 12.2 lists the ten products selected and their corresponding safety screening values. The benchmark of TCE for the safety screening cut off is 37.

Table 12.2. Safety Screen - Comparison of Potential Products

Product	VOC:	GWP:	ODP:	HMIS H:	HMIS F:	HMIS R:	pH:	Safety Screening Score
SC Aircraft and Metal	0	0	0	0	0	0	8.4	49
Polyspray Jet 790 P	0	0	0	1	0	0	10	46
Micro 90	0	0	0	2	0	0	9.7	46
LF 2100	6	0	0	2	0	0	9.5	46
Inproclean 3800	0	0	0	2	0	0	12	42
Valtron 2275	0	0	0	1	1	0	7.3	48
Ozzy Juice SW 1	0	0	0	1	0	0	7	49
Sea Wash Blue	0	0	0	0	0	0	7	50
Ionox HC 2	50	0	0	2	1	0	10.2	42
Aquavantage 1400	0	0	0	1	0	0	11.3	46

Phase II Temperature and Concentration Studies

These products were diluted to 5% using DI water in 600 ml beakers. The dilutions were heated to 130 F on a hot plate. Pre-weighed alloy coupons were coated with the first supplied lubricant and weighed a second time to determine the amount of contaminant added to each coupon. Three coupons were immersed into each solution and cleaned for five minutes using stir-bar agitation. Coupons were rinsed in room temperature tap water for 15 seconds and air dried using compressed dry air for 30 seconds. A final weight was recorded and efficiencies were calculated for each coupon cleaned. Only three products, listed in Table 12.3 were found to have marginal success (approximately 40%) at removing the first lubricant.

Table 12.3. Soil 1 Removal Results

Product	Average Removal
Micro 90	48.13
Ionox HC 2	42.74
Aquavantage 1400	38.50

Follow-up testing with Micro 90, Ionox HC 2, and Aquavantage were conducted at higher concentrations (10%) in an attempt to improve efficiencies. In addition, six non-aqueous products were selected for evaluation as well and are listed in Table. 12.4. One of these products was diluted to 10%. The other five products were used at full strength. The cleaning solutions were all heated to 130 F on a hot plate.

Table 12.4. Non-Aqueous Products Selected

Products	VOC	GWP	ODP	HMIS H	HMIS F	HMIS R	pH	Safety Screening Score
SC Supersolve	0	0	0	0	0	0	10.2	47
Soy Gold 1100	41	0	0	0	1	0	7	48
Solsafe 245	840	0	0	2	1	0	NA	47
Shopmaster RC	1000	0	0	2	1	0	NA	47
DS 108	890	0	0	1	2	0	NA	47
D Greeze 500 LO	450	0	0	0	2	0	NA	38

All five products used at full strength removed over 75% of the lubricant. The three products from the previous trial did not improve with an increase in concentration. Table 12.5 lists the results of cleaning for the second attempt on the first soil.

Table 12.5. Soil 1 Removal Rates

Cleaner	Average Removal
Micro 90 -10%	41.49
Ionox HC 2 -10%	42.13
Aquavantage 1400 -10%	29.63
SC Supersolve -10%; 100%	38.53; 39.18
Soy Gold 1100	83.28
Solsafe 245	97.91
Shopmaster RC	93.77
DS 108	75.36
D Greeze 500 LO	93.78

Having identified several products that were effective on the first lubricant, subsequent testing was conducted on the other soils. Only one soil was difficult for the non-aqueous products to remove. With the addition of a wipe, two of the products removed about 80% of the soil.

Table 12.6. Soil Removal Summary

Product	#2 Soil	#3 Soil	#4 Soil	#4 with wipe	#5 soil
Soy Gold 1100	97.57	91.74	-56.42	NT	95.31
Solsafe 245	99.47	95.20	-36.42	NT	95.41
Shopmaster RC	99.44	95.50	35.46	81.64	92.08
DS 108	99.14	97.15	-42.75	78.62	97.33
D Greeze 500 LO	96.58	97.13	-102.26	NT	94.30
SC Supersolve	71.48	NT			

Phase III Mechanical Energy Studies

The cleaning effectiveness for all products would be improved with the adjustments of the TACT variables, temperature, agitation (ultrasonics or spray washing), and/or time. As was shown on the fourth soil, the addition of mechanical energy (wiping) significantly improved two products in their soil removal.

Phase IV Parts Pre-Pilot Cleaning Studies

The five successful products from the previous trials were selected to test on the supplied parts. Each product was heated to 130° F on a hot plate. The heated solutions were then poured into a tall cylinder (25 ml glass graduated cylinder) and placed in a heated ultrasonic tank. Each solution was degassed for five minutes.

Table 12.7. Final EHS Comparison (P2OASys)

Category	Soy Gold 1100	Solsafe 245	Shopmaster RC	DS 108	D Greeze 500 LO	TCE	Ensolv
Acute Human Effects	2	4	6	5	5	9	6
Chronic Human Effects	2	3	2	2	2	7	6
Physical Hazards	2	4		6	4	7	9
Aquatic Hazard				4		7	2
Persistence/Bioaccumulation	2					9	6
Atmospheric Hazard	2	2	2	2	2	6	10
Disposal Hazard	3	5	6	4	4	8	6
Chemical Hazard	5	10	5	8	6	10	10
Energy/Resource Use	2	4	2	5	3	8	8
Product Hazard	4	5		6	3	7	6
Exposure Potential	2	4		4	2	8	6
Final	26	41	23	46	31	86	75
Weighted Final	2.60	4.56	3.83	4.60	3.44	7.82	6.82
Safety Screening Score	48	47	47	47	38	37	24

Two dirty parts were immersed into the solutions. The parts were placed in the upright position and cleaned for five minutes using a Crest 40 kHz ultrasonic tank at 130 F. Parts were then rinsed in a DI Spray for 15 seconds at room temperature and dried with compressed air for 30 seconds also at room temperature. Visually, all five products removed exterior soils. During cleaning, bubbles were observed coming out the tops of each of the parts which would signify that the cleaning solutions were getting into and out of the inner diameter of the parts. Also noted, when the parts were being packaged, one of the five products – D Greeze 500 LO – left a tacky residue on the parts. Parts were packaged to be sent back to client for final observations. Having identified five possible alternatives through the laboratory testing, a final environmental safety comparison was conducted on the products. All five of the laboratory selected alternatives were assessed with a safer overall EHS profile when using the P2OASys tool. TCE was included with the alternatives in the comparison. As shown in Table 12.7 the Soy Gold

1000 had the best score followed by D Greeze 500 Lo and Shopmaster RC. An additional solvent was included in the comparison to show how the potential drop-in solvent compared to the current solvent.

Phase V Pilot Plant/Scale-up Feasibility Studies

Upon receipt of the parts on-site, the company conducted an internal review of the laboratory cleaned tubes. Despite having promising results on the soils, the final decision to go to a new system (ultrasonic) and abandon their current process (vapor degreasing) was not made. However, having seen that possible alternatives could work did encourage the company to continue the pursuit for other drop-in solvents. A series of internal trials resulted in the selection of a non-regulated (at the time 2007) solvent. They choose nPB for all in-process cleaning and it has worked well for cleaning. Admittedly, VOC emissions however have remained a challenge for them. On a very limited basis, the company tested super-critical carbon dioxide which worked very well for small scale testing. However, scaling-up a facility would be too cost prohibitive.

Conclusion/Next Steps

There were several products that worked well on nearly all of the supplied contaminants using immersion cleaning in the laboratory setting. Additional testing was recommended to determine the optimum conditions for removing soils four and five. Based on the company's need for using existing equipment and bolstered by the possibility of alternatives existing, the company completed the project by evaluating drop-in solvent on their own.

12.2. Case Study Capacitor Cleaning - Building Capacity to Replace

Summary

Back in 2004, a leading provider of film capacitors for industrial, medical, and specialized applications serving original equipment manufacturers (OEM) and distributors needed to come into compliance with the regulatory requirements at their facility. Their centralized facility included R&D, dielectric development, product design, test labs, automated manufacturing, warehousing, and customer service.

Background and Problem statement

The Capacitor Company had just spent more than a million dollars upgrading their Tiyoda Serec F1 Clean System. It was in place to clean the various surface types including aluminum, tin plated steel, ceramic, and nylon 66 plastic parts. The company was cleaning 50,000 parts a day over two shifts. The improvements of the vacuum degreaser came after they were found to be in noncompliance with their air quality permits. Prior to the upgrades, the company had tried to use aqueous systems. However, these water cleaning processes left trace amounts of residue which were not acceptable to their clients.

So the stage was set for them to find a solvent that would be compatible with their current system and clean the multiple soil types from the capacitor materials all while bringing themselves into compliance with state and federal air regulations.

Soils to be cleaned included four metal working fluids: 1-Soltex Polybutene 32 (9003-29-6); 2-Cargill, Inc Canola Oil (120962-03-0); 3-Nisseki SAS 40 oil (27776-01-8, 612-00-0, 103-29-7, 101-81-5); and 4-C.P. Hall Co. Plasthall ESO oil.

Phase I Product Selection

Four products were selected based on the client's request for vapor degreasing solvents. An additional set of products were later identified and added to the evaluation process. The preliminary safety screening was not used to screen out products for this project due to the request for all available drop-in solvents. Safety scores ranged from 28 to 39. Recall that TCE was scored at a 37.

Table 12.8. Safety Screen - Comparison of Potential Products

Product	VOC	GWP	ODP	HMIS H	HMIS F	HMIS R	pH:	Safety Screening Score
HFE 7100	0	480	0	0	1	0	NA	39
HFE 7200	0	55	0	3	1	0	NA	31
HFE 71 DE	0	320	0	0	1	0	NA	39
AK 225	0	> 1	> 1	1	1	0	NA	33
Vertrel CCA	526	678	0	1	0	1	NA	28
Vertrel MCA	536	806	0	1	0	1	NA	28

Phase II Temperature and Concentration Studies

Ideally, the selected products would be effective at removing all four of the company's soils. To do this, the first round of testing included the soil the company used most often. Aluminum surfaces were coated with the Soltex Polybutene 32 and cleaned in each solution for five minutes using little to no agitation. Four of the six products worked extremely well, removing more than 90% of the soil in five minutes of cleaning. Table 12.9 lists the summary of the testing. All products were kept for the testing on the second soil.

In this evaluation two of the six products were successful in removing more than 85% of the Canola Oil in the five minutes of immersion cleaning. Many of the other products left residual oil on the surface. When looking at the third soil, Nisseki SAS 40 oil, four of the six products removed over 85% of the oil from the aluminum coupons within the five minute of cleaning. During the cleaning of the final soil, Plasthall ESO oil, none of the products worked well within the five minutes of cleaning, removing less than 85% of the soil. Follow-up testing was planned to use longer times. The transition to Phase III would introduce the use of vapor degreasing.

Upon reviewing the results with the company, they expressed an interest to continue identifying other drop-in solvents that would be more consistently successful across all of the soils. Only one of the six was determined to be effective on all four soils. With this request, the lab included three nPB solvents and two additional HFE solvents. The summary of each of these products listed in Table 12.10 includes safety screen score and soil removal for all four soils. All had low safety scores (five of the six under 25 and all under 37). Each of the added products was successful in removing more than 95% of the soil in all but one product and two soils.

Table 12.9. Soil Removal for All Soils

Product	Soil 1	Soil 1 T2	Soil 2	Soil 3	Soil 4
HFE 7100	9.25	14.97	53.58	99.58	52.00
HFE 7200	19.05	39.52	55.64	101.57	58.75
HFE 71DE	97.39	NT	86.46	77.77	80.56
AK 225	94.03	NT	73.28	97.10	84.32
CCA	97.25	98.71	86.38	86.82	84.95
MCA	97.15	98.02	78.13	74.03	75.08

Table 12.10. Added Solvents Safety and Performance

Product	Safety Screen	Soil 1	Soil 2	Soil 3	Soil 4
Ensolv (nPB)	24	99.73	99.62	99.21	99.64
Ensolv A (nPB)	24	99.81	99.74	99.15	98.47
Metalnox M6960 (nPB)	23	97.68	100.01	100.06	99.56
Flux Remover (HFE)	33	99.66	98.42	97.45	98.09
Heavy Duty Degreaser C (HFE)	23	90.31	87.33	95.69	98.29

Phase III Mechanical Energy Studies

One product, Ensolv, was selected based on client input from the previous phase of testing. Cleaning utilized a bench top scale vapor degreaser operating at the boiling temperature of nPB. Coupons were immersed for five minutes in the vapor zone, removed, and cooled. The product showed immediate success with the introduction of the vapor degreasing. Table 12.11 lists the removal rates for the four soils.

Table 12.11. Vapor Degreasing with Ensolv

Soil	Average
1 - Soltex	99.98
2 - Cargill	99.88
3- Nisseki	99.43
4 - CP Hall	99.98

Having identified possible alternatives through the laboratory testing, a final environmental safety comparison was conducted on the products. The laboratory-selected alternatives were assessed with a slightly better overall EHS profile when using the P2OASys tool as compared to TCE. Table 12.12

lists the comparison of the two solvents. Despite having a lower safety screening score, nPB would provide the company with a small upgrade in EHS concerns.

Table 12.12. EHS Comparison of Selected Solvent

Category	Ensolv	TCE
Acute Human Effects	6	9
Chronic Human Effects	6	7
Physical Hazards	9	7
Aquatic Hazard	2	7
Persistence/Bioaccumulation	6	9
Atmospheric Hazard	10	6
Disposal Hazard	6	8
Chemical Hazard	10	10
Energy/Resource Use	8	8
Product Hazard	6	7
Exposure Potential	6	8
Final	75	86
Weighted Final	6.82	7.82
Safety Screening Score	24	37

Phase IV Actual Product Cleaning Studies

Due to the makeup of the company's parts, testing the solvent on supplied parts in the lab was not feasible. Piloting on-site was proposed as the next phase of the project.

Phase V Pilot Plant / Scale-up Feasibility Studies

Ensolv was selected by the company for in-house testing. They contacted the vendor to have a bench top pilot vapor degreaser brought in for the piloting. After two months of piloting and trouble shooting, the company replaced their current solvent with the nPB solvent.

Conclusions Project 1/Next Steps

One might think this was the end of the story. Remember, the main reason the company chose to go with a drop-in solvent was that they had invested heavily in their current equipment and could not justify the cost to replace it with a new system.

Fast forward 10 years. Regulations change, equipment gets used, becomes old and has wear and tear. The company is now looking to replace the equipment and related solvent. Enter round two of the project.

Project 2 Same Place, Different time, Same Bat Channel – The project continues!

Summary

The Capacitor Company who left PCE and TCE over a decade ago for nPB resurfaced and asked the TURI Cleaning Lab for help finding equipment and cleaners that could replace their now old, custom made, vacuum vapor degreasing machine. N Propyl bromide, the most popular drop-in replacement from the mid 2000's for TCE and PCE in vapor degreasing, was now being regulated in Massachusetts and was slated to become a high hazard substance, HHS soon. This meant that the reporting threshold to the state would now be lowered to 1000lbs or about a drum and a half. The company's use would be much higher than this lower threshold and they wanted to be proactive to replace the solvent.

Background and Problem Statement

The main push to work on a new cleaning system was that the company's custom, vacuum, vapor degreasing system was getting old and costly to maintain. So before replacement became a desperate situation or a major breakdown happened and affected production, they wanted to start looking at options. They wanted to do toxic use reduction and move away from a system that used toxic solvents to clean their capacitors.

This company had been a large manufacturer in the state of Massachusetts for years and had been working on site cleanup with the EPA for years which further motivated the company to move toward a cleaning system that would protect their workers, the environment, and the community in which they did business.

They wanted to retain as much of their current baskets and conveyor belt system as they could to deliver the parts during cleaning. They were open to go semi-aqueous or full aqueous if possible in order to leave nPB. As part of the process they were investigating ways to utilize evaporation and water recycling systems. The company also wanted a system that would give them flexibility and options in the future if they added any new products to their production.

So the water cleaning process that was not an option over a decade before was now viewed as an acceptable method for them and their clients. Therefore, the stage was set for them to find an aqueous or semi-aqueous system that would be compatible with their current production process, cleaning multiple soils and metals used on the production floor. The soils to be cleaned included the same four metal working fluids identified during the first project: 1-Soltex Polybutene 32 (9003-29-6); 2-Cargill, Inc Canola Oil (120962-03-0); 3-Nisseki SAS 40 oil (27776-01-8, 612-00-0, 103-29-7, 101-81-5); and 4-C.P. Hall Co. Plasthall ESO oil.

Phase I – Cleaning System and Equipment Selection

For the first step in the second round of the project, product selection was very important. Since the way the cleaner was going to be applied to the parts, how the parts would be rinsed, and the way drying was going to be done were to be changed, the selection of a cleaner was more difficult in the second project because it would need to be intimately tied to the new machinery. Both things had to be researched and tested concurrently. Everything was going to be started anew, not just the chemical.

In this phase, water was added to the system resulting in surface tension of the cleaning system being higher, creating the potential for the cleaning solution or solutions to stick or stay in corners and any blind holes in the parts. This added a complication of the parts rusting or oxidizing if they were not rinsed and dried correctly prior to prolonged storage. To make sure this did not happen, a lot of testing would need to be done with the company's parts with any potential cleaning systems in order to rule out any of these issues. The main testing would focus on the least robust surface, like regular steel, which can flash rust and oxidize, and any softer metals like aluminum to make sure the new solutions did not damage the metals.

As mentioned, the cleaning step would no longer be the only phase of the system that could be a potential problem. Now, the process would need a rinse phase of the system, which could cause an issue. Wetting agents would be looked at to avoid any issues. In the new drying phase, the addition of compressed air would be evaluated and if that did not work then nitrogen or another gas may be used. The good news was that these parts were not needed to be microscopically spot free and so regular compressed air should work for their needs.

Dialogue with the company about the new system began by discussing the concepts of using immersion cleaning, ultrasonic cleaning, and spray cleaning. In each scenario, the use of a conveyor belt or an automated lift system was posed as possible options. Variables abounded for to the new cleaning system, as well as equipment. The system would be of significant expense to buy, custom make, install, and retrofit the facility to handle it. Fortunately, the company had good space and infrastructure for the room sized machinery.

Phase II – Contacting Vendors and Technical Resources

After deciding that the same vapor degreasing equipment would not be used this time, the next step was to talk to vendors and other technical experts in the metal cleaning field about the potential processes and chemicals to be used on the company's capacitors. To further complicate the selection of any new cleaning system, the company revealed that some of the capacitors were cleaned only on the outside and others that were made in-house needed to be cleaned inside before being impregnated with the oils that conduct the electricity for their use. In the latter case, the cleaners and any other residue remaining on the inside of the capacitor had to be tested for any possible interference in conduction.

Expertise was sought from local vendors of chemicals and equipment. In addition, other metal finishers using similar cleaning systems were queried. Further input was solicited about the major metals being used, including stainless steel, steel, and aluminum from the local surface finishing trade group. The consensus was that a phosphating cleaning system should be used, as this would not only clean the oils and the waxes but would protect the regular steel from rusting or oxidizing. The remaining question was whether the phosphating system would interfere with the oil used in the capacitors.

With the type of chemistry to use identified, the focus shifted to identifying the equipment. The equipment options were narrowed down based on the company's needs. The investigation was to look at spray cleaning with a conveyor belt and a cabinet type or immersion tank on an automated line. The group's feeling was that a spray line would work best to clean all areas of the capacitors while utilizing both the current conveyor belts and the existing delivery baskets from the old system.

Phase III –Chemistry in the Lab and in the Field

The next step was to try phosphate cleaners on the soils to see if effective removal using basic immersion cleaning at room temperature was possible. Other variables, including, heat, mechanical energy, and concentration would eventually be worked on during piloting.

In the laboratory, testing was done using room temperature immersion cleaning with a 5% solution on the substrates of steel, stainless steel, and aluminum. The cleaning solutions showed effectiveness under these basic parameters. However, the true test would to be to take the worst case dirty capacitors and clean them. These would be the capacitors that needed to be cleaned on the outside and those units that needed the insides cleaned. Testing was set up to run both types through a cleaning line using the lab tested solution.

A company using the same cleaner was located through a vendor that was willing to offer up their facility for piloting. Their equipment was a bit different than the one being researched for the client. The testing line at the test location had a large spray line where the parts flowed through it on hooks. Also, the system cleaned large parts versus the smaller capacitors that would be in baskets in the client's system. This was only seen as a minor difference and testing proceeded. Three baskets of differing types of capacitors from the client were run through the line and sent back to the client for analysis.

The analysis was done by the client's in-house engineers to see if the proposed cleaner left a residue that would interfere with how the capacitor worked. Validation, which would take up to six months, also included checking the other capacitors that were cleaned while closed to make sure they were cleaned sufficiently.

Phase IV – Equipment Research While Waiting for Client Testing

While the client tested the cleaned capacitors, the team of technical experts researched the equipment that would be the best for this location and client. The vendor was known to work well with customers, was a nation-wide company, but had great local representatives to help install the equipment as well as train the client on how to operate the system. Since equipment and chemistry have to work together, the equipment company was asked to locate one of their customers that was going to be using a spray conveyor belt machine with a phosphating cleaner. This connection would allow for the client to discuss any reservations or get implementation questions answered.

In the meantime, the team checked in with the client to see how quality control testing was going. The report from the client's engineers showed three positive outcomes. First was that the residue from the cleaner was stopping any flash rusting or oxidation on the steel capacitors. Secondly, the process did not damage the aluminum. Finally, the cleaning did not have any effect on the oils in the capacitors that were washed before filling and closing them. This was a huge hurdle overcome.

From there, the next step reverted back to finding a line that was as similar to the proposed system as possible. If the exact chemistry could not be found with the right equipment, then two options existed. The first would be for the equipment company to pilot the exact chemistry in as close of a system as possible. Option two would be to see if another company with the exact line would allow the specific chemistry to be run in their machine for a pilot. All of the options were possible, but the logistics would need to be worked out.

Phase V – Murphy's Law – What Can Happen Will Happen

With the finish line in site and many of the challenges conquered, Mr. Murphy walked in. The company put the brakes on, pulled the parachute, and took a detour, walked away or however you want to say it. As everyone knows, business is unpredictable and sometimes during a project things come up that the company has to concentrate on more than others. The team was notified by the operations officer that the company had other more pressing projects going on that they had to divert their attention to at this time. They were appreciative of the work done by all parties involved and promised they would be back. Based on their track record, they were to be believed.

All members of the team were contacted and told the project was on hold, and while disappointed, it was not anything anyone on the team had not dealt with before. It cannot be taken personally. The positive outcome was not tied to the adoption of a system, but to the good work and trust that was built up over the course of the project.

Conclusions/Next Steps

The next steps when a project is put on hold are to keep in periodic contact with the client but not hound them. They have a right to do what they want/need and it is the team's job to stay available as a resource. Knowing your place in the client's life is paramount. An email every quarter is what is called for here, and when interest is shown, the frequency can be increased, but the client is the one who sets the pace. As a free technical service in Massachusetts, the TURI Cleaning Lab is aware that it is working for the client. It is always a privilege to be welcomed into a company to work with them and their team. To be asked back a second time is an honor.

Build the trust and they will return!

12.3. Case Study Collaborative Efforts to TCE Reduction: Don't Go It Alone

A successful method for providing technical assistance for TCE reduction in the state of Massachusetts has been established and refined over the years. This process has benefitted the companies that can easily access the services. The effort has grown out of the collaboration between various state agencies focusing on pollution prevention initiatives. During the middle of the last decade, this model of combined assistance was expanded in an effort to help another state tackle their growing problem with TCE emissions.

For many cases, particularly among larger companies, uses and emissions of TCE have been minimized through substitution and other engineering controls. Unfortunately, the use of TCE has not been eliminated. Alarmingly, there are many small users of TCE that are utilizing the solvent for cleaning practices. These small companies are typically located in older buildings and in some instances within close proximity to residential dwellings. As mentioned, the equipment used by the facilities is not state-of-the-art and contribute to higher levels of TCE in the area. Figures 12.2-12.6 depict representative facilities and equipment used in the Providence, Rhode Island area.

Figure 12.2. Mill Building Location.
Photograph courtesy of Linda Darveau.

Figure 12.3. Interior Building Layout.
Photograph courtesy of Linda Darveau.

Figure 12.4. Facility located adjacent to Residential Housing.
Photograph courtesy of Linda Darveau.

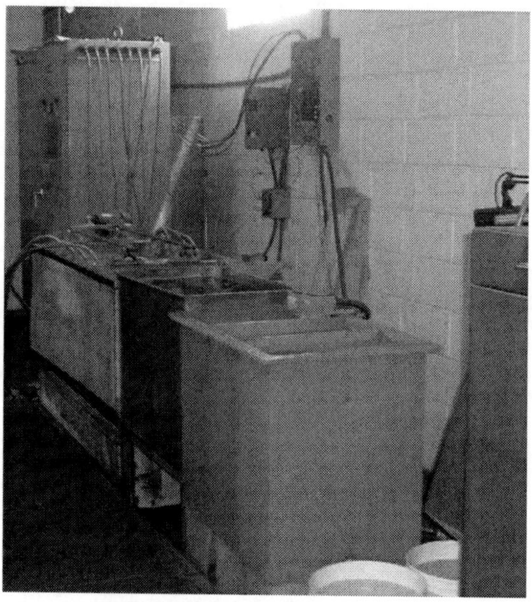

Figure 12.5. Degreasing Equipment.
Photograph courtesy of Linda Darveau.

EPA Region 1 found TCE at elevated levels in the Providence, Rhode Island area as part of EPA's routine air monitoring system evaluation. Following EPA's environmental justice efforts, Region 1 made TCE the focus area over the next year. Using their resources, they generated a list of approximately 40 companies potentially using TCE. This list was then visited by EPA Region 1 staff members. Many of the companies that were identified made materials for the Department of Defense (DoD). When the DoD learned of TCE still being used by its suppliers, they encouraged these suppliers to eliminate the use of the chemical.

Figure 12.6. Open Top Degreaser.
Photograph courtesy of Linda Darveau.

With the list of companies in hand, the EPA then partnered with the Rhode Island Department of Environmental Management (RI DEM). They were tasked with figuring out a way to eliminate/reduce the usage of TCE in this community. Enforcement actions were considered, but the regulatory partners agreed that a more effective method for improving air quality in the communities affected should be researched and offered. EPA Region 1 connected with the TURI Cleaning Lab to assist in providing technical assistance and/or training to the identified companies. The end goal was to provide support to the companies to voluntarily reduce/eliminate the use of TCE.

The collaboration began with EPA Region 1 personnel contacting the companies from the list to collect dirty parts and soils that would be used for laboratory research efforts. Testing was performed on several alternatives that were appropriate for each company's needs. With the list of possible products ready, companies were contacted with the results.

Follow-up assistance on a one-to-one basis would be challenging to do on-site. Therefore, the combined efforts of EPA Region 1, TURI Cleaning Lab, and RI DEM were maximized by holding a training workshop. The laboratory proven products and bench scale equipment were made available to the individual companies to conduct "on-site" piloting, and the TURI Cleaning Lab and company workers were jointly responsible for the cleaning process. An added benefit was opening up lines of communication relative to what actually goes on in a process of this size. Various companies realized they were not alone as dialogue with neighboring companies revealed that similar concerns and obstacles existed. By working together, many facilities were more willing to begin to adopt safer alternatives. The major driving force was the DoD's encouragement (not quite an outright demand) to replace TCE. If one company was going to do it, the others recognized the need for them to join in the effort or be left out of the DoD supply chain.

A second hands-on training workshop was held two years later with the target audience being those companies that could not attend the initial workshop. This new training was modified from the original to provide the lessons learned by the various partnering organizations. The materials provided included a tip sheet generated by the joint effort of EPA Region 1 and TURI Cleaning Lab and short case studies of some of the companies that had successfully substituted TCE. These companies provided personal contact information to other companies that may want to discuss the project in person.

The combined efforts of EPA Headquarters, EPA Region 1, TURI Cleaning Lab, RI DEM, and the Narragansett Bay Commission (NBC) (which joined the group to help with water discharge issues) set out to contact companies from the original list provided by EPA. In addition, the NBC sent a copy of the flyer to more than 80 companies that the NBC thought would benefit from the workshop. A handful of Massachusetts companies, identified by the Massachusetts Office of Technical Assistance and located close to Rhode Island, also were invited to the conference. In total, there were eight companies that signed up for the workshop, six from Rhode Island and two from Massachusetts. In addition to TURI Cleaning Lab staff and EPA Region 1 contacts, representatives from the RI DEM, NBC, and Rhode Island Occupational Safety and Health Agency (RI OSHA) attended the workshop and were available for companies to discuss how each agency could assist companies in TCE replacement/reduction.

The goal of the project was to reduce the amount of TCE used by small companies in Rhode Island. During the period after conducting the workshops, the TURI Cleaning Lab had direct contact with a dozen companies. The amount of TCE that was being used by attendees and other contacted companies was calculated to be about 26,000 pounds per year. This number was based on the preliminary questionnaire and/or registration forms for new companies serviced and those that continued during the second year of work.

Services provided to companies included preliminary site visits by TURI Cleaning Lab personnel and EPA Region 1 staff, laboratory testing, follow-up site visits and on-site testing assistance. The visits allowed the TURI Cleaning Lab to conduct a walkthrough of the facility and to collect sample materials for laboratory testing.

Through the combined efforts of all of the agencies and businesses involved, the companies were successful in eliminating 82% of the reported TCE usage, a combined reduction from the 26,000 pounds per year to less than 7,000 pounds per year. Additionally, the process helped to reduce the use of another solvent, n-propyl bromide (nPB) for one company. This company had reported their initial usage of nPB at 6,000 pounds per year, and after working with the various agencies, the company was able to reduce usage to well below 1,000 pounds per year.

Summary

Taking one state's resource and combining it with other regional and national organizations resulted in a dramatic and efficient reduction in the use of TCE for solvent cleaning. Partners in this project included EPA Region 1, EPA Headquarters, DoD, RI DEM, RI OSHA, NBC, Massachusetts Office of Technical Assistance, and the TURI Cleaning Lab.

The lessons learned for TCE users in Rhode Island can easily be applied to other areas with concentrated industry regions. Connecticut may be one such region with localized job shops and metal working facilities. According to EPA sources, Bridgeport, Danbury, and New Haven may be ideal localized areas to provide future hands-on workshops and on-site cleaning assistance by the TURI Cleaning Lab or a similar operation. The methodology for on-site assistance follow-up to the preexisting TURI Cleaning Lab testing program has been incorporated into the work the TURI Cleaning Lab conducts for companies in Massachusetts.

Chapter 13

FUTURE/EMERGING SOLVENTS/NEXT STEPS

ABSTRACT

TCE has excellent solvency, works to clean and protect parts for the next manufacturing step, storage or shipping. It is also the "devil you know" for many shops, but it is very toxic to the environment and human health. There are many viable options if a company is willing to put the time, effort and money into making the switch. This chapter puts forth a way to accelerate the transference of the TURI Cleaning Lab blue print and its localized success in reducing TCE in New England to the rest of the country.

PROPOSED WORK FOR TCE SUBSTITUTION

As we mentioned earlier, many people believe that TCE is an old technology and no one uses it anymore. But the Toxics Release Inventory (TRI) data and the work done in New England show that TCE is still being used. As a cleaner, TCE has excellent solvency and works to clean and protect parts for the next manufacturing step, storage, or shipping. It is also the "devil you know" for many shops, but it is very toxic to the environment and human health, and there are many viable alternatives if a company is willing to put the time, effort, and money into making the switch.

Consultants, state, and federal pollution prevention professionals can help companies make this switch. What also resonates with TCE users in specific sectors is learning about other companies in their sector making the switch. Hearing and seeing how the process has worked goes a long way to making the case for change. This can be in the form of case studies, seminars, webinars, published articles, and trade show educational sessions.

Information may be the key to moving reluctant or overwhelmed companies away from TCE, but money, time to research, and testing for client requirements is also needed. Below is a proposed collaboration between the TURI Cleaning Lab, EPA Headquarters, EPA's regional offices, and state pollution prevention organizations. In Chapter 12, case studies of field and lab work done to help companies make the solvent switch were discussed. This future work is a collaborative effort to work on a larger project in order to document and show how the switch from TCE can be done. This work will help the EPA on future regulations and also help disseminate technical information to those who use TCE.

The challenge with this type of work is how to reach the small, underground or non-registered businesses in the country using minimal amounts of TCE in shops that go unregulated. Past experience has shown that these shops are run by people who have worked previously at larger shops and opened their own small shops than handle overflow work. These companies may rent a small space or work in a home shop and may or may not register in the city or town in which they are located. Due to their invisibility cloak, these shops may not have connections to local and national trade associations or state and federal agencies that regulate chemistry in their sector. In addition, and more importantly, they may not be fully aware of how dangerous the chemicals they are using are to their health and the environment. Disposal procedures also may not be known. Finding and reaching out to this subpopulation that uses TCE has to be a priority for any future work by regulators and pollution prevention groups.

EPA-TURI Cleaning Lab Collaboration

TCE was identified for risk assessment as part of EPA's Existing Chemicals Management Program. EPA reviewed readily available information on TCE, including uses, physical and chemistry properties, fate, exposure potential, and associated hazards to humans and the environment. TCE was selected based on concerns for its human health hazard (e.g., human carcinogen), and its exposure profile (i.e., widely used in consumer products and detected in drinking water, indoor environments, surface water, ambient air, groundwater, and soil) using the Office of Pollution Prevention and Toxics Substances Control Act (OPPT's TSCA) Work Plan screening methodology.

In January 2013, EPA released the draft risk assessment on TCE for public comment and peer review. EPA then addressed the comments obtained through public and peer review, and released the TCE risk assessment in May 2014. This risk assessment evaluated the human health hazards (i.e., non-cancer and cancer) from inhalation exposures to hobbyists/consumers where TCE is used as a clear protective coating spray, to workers and hobbyists/consumers where TCE is used as a degreaser, and TCE use as a spotting agent in dry cleaning. EPA's preliminary conclusions indicate that the risks to consumers and workers associated with these TCE uses are significant.

An expert public workshop on alternatives and risk reduction approaches to TCE was held during the summer of 2014. The goals of the workshop were to generate information from experts on TCE when used as a degreaser, better understand the factors when transitioning to safer alternatives, and to develop options to address the risks identified in the assessment.

The workshop collected more information on how TCE is used as a degreaser, understand the process and efficacy considerations important to selecting safer alternatives, solicit feedback on potential alternatives to TCE, and discuss safer engineering practices and technologies that can reduce exposure to TCE. Information from the workshop will be helpful as EPA works with stakeholders on potential risk reduction approaches.

EPA is pursuing both voluntary (labeling, phase-out of consumer uses) and regulatory (Section 6 and Significantly New Uses Rule SNURs) options for the uses identified in the risk assessment. There were some other uses of TCE not in the risk assessment, so more analyses are being conducted to determine if EPA should include those uses.

Planned Work

In the spring of 2015, the EPA issued a Statement of Work to support testing to assist TCE users in identified metal refinishing locations in the United States to switch from using TCE to safer, less toxic alternative cleaning solutions. The documented results of this testing will be for use by EPA to assist in further voluntary and regulatory actions. The work will build off of similar projects the TURI Cleaning Lab has conducted.

To begin, using EPA's TRI database, the TURI Cleaning Lab will locate metal refinishing facilities in various locations throughout the U.S. in states where TCE is still used. The contractor will submit the recommended locations to EPA for approval. With EPA concurrence, the locations will be selected where the studies will be performed. From the TRI database, TCE was reported in 37 states during 2013. Table 13.1 lists the states reporting TCE in 2013.

Table 13.1. States Reporting TCE Disposal or Other Releases for 2013

Location	Disposal/Releases in Pounds
Arizona	14033
Arkansas	5,271
California	260
Colorado	16,292
Connecticut	65,967
Delaware	6,046
Florida	13,870
Georgia	15,090
Illinois	105,548
Indiana	246,723
Iowa	11,223
Kansas	131,952
Kentucky	28,351
Louisiana	44,424
Maryland	71,794
Massachusetts	28,092
Michigan	35,492
Minnesota	126,904
Missouri	63,875
Nebraska	21,307
Nevada	500
New Jersey	14,864
New Mexico	108,711
New York	53,287
Ohio	120,649
Oklahoma	13,040
Oregon	184,239
Pennsylvania	147,262
Rhode Island	10
South Carolina	8,877
Tennessee	162,703
Texas	28,904
Utah	8,867
Vermont	1,546
Virginia	25,952
Washington	10,184
Wisconsin	6,634

The TURI Cleaning Lab initially looked at the top nine states (all over 100,000 pounds TCE). From there the lab investigated the companies reporting TCE in these states and categorized the sectors. These sectors are listed in Table 13.2. The top sector was acid-battery manufacturing, which is outside the scope of the project and expertise of the lab. The major areas included tube manufacturing in Pennsylvania, metal working in Illinois and Minnesota, and aircraft manufacturing in Kansas.

Table 13.2. Top Industries Reporting TCE Disposal in 2013

Report Disposal (lbs)	Industry	Notes
510,821	Acid-Battery Manufacturing	
131,940	Tube Manufacturing	All located in Pennsylvania
163,134	Metal Working	41% in Illinois, 31% Minnesota
122,946	Aircraft	Most in Kansas (~90%)
39,961	Medical Instruments and Devices	
70,669	Chemical Distributor, Manufacturing	
124,650	Waste Handling	

To best utilize time and resources, when possible the TURI Cleaning Lab will work with companies that are centered geographically. For Pennsylvania, a couple of the companies are too far away from the others and would make it hard to include in one trip, but two of them are close to each other and are also the larger users of TCE. The aircraft manufacturing companies were within a twelve mile radius and could all be included. The metal working in Minnesota could include three of the five facilities and account for 70% of the TCE usage by this sector in the state while staying within a 35 mile radius. In Illinois, five of the six companies could be serviced within a 30 mile radius.

These companies that have been identified as possible sites for future work make up about 30% of the total TCE reported and 50% of the TCE when omitting the acid-battery sector. The companies as described above would be about 70% of the reported TCE for the three sectors. Tables 13.3 through13.6 list the company locations and amount of TCE reported by each in 2013.

Table 13.3. Pennsylvania Targeted Areas Using TCE [51]

Tube Manufacturing		
City	TRAPPE	KING OF PRUSSIA
TCE reported (lbs)	68,924	5,940
Main work focus	Iron and Steel Pipe and Tube Manufacturing from Purchased Steel	Producing high quality tubing
Local P2 Resource	Pennsylvania Office of Pollution Prevention & Energy Assistance http://www.portal.state.pa.us/	

Table 13.4. Kansas Targeted Areas Using TCE [51]

Aircraft Mfr				
City	WICHITA	WICHITA	WICHITA	WICHITA
TCE reported	24,031	33,000	57,415	18,330
Main work focus	Working in the Metal Finishing industry	The leading designer and manufacturer of light and midsize business jets, utility turboprops and single engine aircraft.	Produce large aero-structures and components of the world's premier aircraft.	Fabricate the sheet metal parts and assemblies for aircraft components
Local P2 Resource	K-State Pollution Prevention Institute http://www.sbeap.org/			

Table 13.5. Minnesota Targeted Areas Using TCE [51]

Metal Working in MN			
City	WHITE BEAR LAKE	SAINT PAUL	MINNEAPOLIS
TCE reported	17,496	15,566	8,518
Main work focus	Supplier of innovative products to the recreational fishing markets. Initial products included fishing lures and lead sinkers. Includes industry-leading terminals for lead-acid batteries.	Manufactured tools	Metal Finishing Industry
Local P2 Resource	Minnesota Technical Assistance Program http://www.mntap.umn.edu/		

Table 13.6. Targeted Areas Using TCE [51]

Metal Working IL					
City	ALGONQUIN	BENSENVILLE	SCHILLER PARK	PALATINE	SCHAUMBURG
TCE reported	7,389	10,088	17,100	10,296	7,840
Main work focus	Intricate metal stampings for the automotive, electronic, communication, electrical-mechanical and medical markets.	Metal forming, machining & fabrication	Tool and die shop	Metal finishing services to manufacturers needing decorative and functional electroplating	Plating and Finishing specializing in: High Phosphorous Electroless Nickel and Electroless Nickel w/ Teflon
Local P2 Resource	Great Lakes Regional Pollution Prevention Roundtable Illinois Sustainable Technology Center http://www.glrppr.org/				

To best conduct on-site visits, the TURI Cleaning Lab would conduct a week-long training for one to two representatives from each of the four regions. These representatives would be someone from the P2 program and someone from the corresponding EPA Regional office. During the training, the lab would demonstrate product selection process, laboratory performance testing, alternatives assessment of products, and site visit walk through with a representative company from Massachusetts. This would serve as an introduction to the process and give context on how the TURI Cleaning Lab integrates laboratory and field work.

Prior to the training, the list of potential companies will be shared with the local P2 programs and an outreach effort will be made to get the companies onboard with the project. As part of the process, a commitment letter should be obtained from the companies before the on-site visits are scheduled. Outreach will also include phone calls or emails to talk to each company about the project and what will be done.

While on location for each of the proposed regions, the TURI Cleaning Lab will meet for four to eight hours with the identified P2 organization and EPA regional personnel to reconnect and prepare for the company visits. The following day the TURI Cleaning Lab representatives and the local P2 designee and EPA regional representative, if desired, will begin the initial onsite visit with the local companies. By incorporating the local P2 and EPA program, future visits to other facilities outside the initial scope of the project in their region can be conducted without the presence of the TURI Cleaning Lab. Essentially, a train-the-trainer program will be established during the work of the project. Each facility will have a four to six hour visit and a one to two hour debriefing meeting between TURI Cleaning Lab and P2 staff.

As part of the lab's testing methods, an initial walk through of each facility will be needed to ensure adequate understanding of the cleaning process, obtain input from workers conducting the cleaning, and arrange for collection of sample parts and soils. These samples will be sent back to the TURI Cleaning Lab to conduct performance testing under the identified and documented real-world conditions. Each region would require up to two weeks of on-site visits.

Using the lab's CleanerSolutions database, a list of alternatives generated from the database will help companies select products that will most likely work for each specific situation. The next step will be to use the operating parameters of temperature, agitation, chemical action, and time, collectively known as TACT, to develop an appropriate and effective cleaning process. At the end of this testing, the more effective products will be retained for the next set of experiments. Next, cleaning will be concerned with the significance of part shape (screw configurations and blind holes, et al) on cleaning. Note that in this model, mechanical energies (Phase III) are critically examined only after the proper cleaners, concentrations, and temperatures (Phase II) are achieved. Laboratory testing will require two months per region. The TURI Cleaning Lab may be able to overlap the testing from the various regions for a total of four to six months of lab testing.

After laboratory testing is concluded for each of the project companies, the TURI Cleaning Lab will again partner with the local contacts. A half-day meeting will be set up with the service providers to discuss a plan of action for each of the companies (most of this can be done via phone and email prior to travel).

Upon successful completion of laboratory testing on collected soils and parts, the TURI Cleaning Lab will return to the facilities to pilot the top two to three products with workers. When applicable, the lab will arrange to have chemistry and bench scale equipment available for the facility if needed. The visit will take between eight and 16 hours. For the visits that will take longer than one day, a return visit later in the week will be scheduled. This will allow the company to work with the piloting equipment and chemistry set-up for a day or two on their own. During this time, the lab will move on to the next company in the region to begin their piloting set-up.

The chemistry and equipment will be left on-site for a minimum of two months. During this time, follow-up communication will be conducted with both the company and the local P2 program.

The TURI Cleaning Lab will conduct follow-up laboratory testing to address any problems that may surface during the pilot testing. During this timeframe, the lab also will help coordinate interactions with the company and vendor(s). For this phase of assistance, projects may need four to 12 months of internal testing depending on the complexity of the project (i.e., company, client, and customer needs for accepting a new cleaning system), resources available, and commitment from the company. A follow-up visit by the TURI Cleaning Lab to the regional P2 offices will be conducted and could possibly include a short (one to two hours) return visit to the companies.

A final conference call with all the regions will be held to discuss the project and future endeavors to provide services to other companies in their regions. A webinar documenting the process may also be an option to help facilitate the work into other areas of the country. Additionally, a train-the-trainer protocol could be established to help other states and EPA regions provide more specialized and personal technical and field assistance to their companies in need.

Model for Other Regions

Depending on the success of the proposed training, site visiting, lab testing, and piloting in the various regions, the proposed TCE reduction project could be expanded to other regions and industrial sectors still using TCE and other toxic solvents in cleaning applications. The end infrastructure of regionally trained P2 personnel will accelerate the transference of the TURI Cleaning Lab blueprint and its localized success in reducing TCE in Massachusetts and New England.

So This Means We have Solved the Problem of How to Move Away from TCE, Right?

While this project will be a large effort around TCE reduction centered on cleaning, it is not the Silver Bullet. What is known about this work is that the soils, parts, and shops may be similar, but each company that uses a toxic solvent like TCE to clean has its own unique challenges and needs. Many of these needs are centered on the facility, management, capital, time, and sometimes most important, the customers that require substantial testing to approve a cleaning technology switch.

These challenges and obstacles can be overcome with patience, work, technical help, and a desire to do the right thing for the environment, workers, clients, and the community in which a company operates. Many times this work is best done in bite-sized pieces or as an iterative process. That way any mistakes are manageable to fix and no one gets overwhelmed with the process.

Good luck to all and remember, no one is alone; there is help available if you need and want it.

Chapter 14

Summary: Ending the Search for the Silver Bullet

Abstract

Since 1993, the TURI Lab has been in search of a quick and effective solution to the TCE substitution problem and has learned that one does not exist. In the absence of the elusive TCE silver bullet, the Cleaning Lab's systematic approach has been optimized and utilized to aid in this search for replacements, drop-ins or process changes to cleaning challenges.

Ending the Search for the Silver Bullet, Embracing the Silver Lining

TCE is a highly versatile solvent that works on a lot of soils and end user applications. TCE is an effective solvent for a variety of organic materials. The solvent is also easy to use and relatively inexpensive. For this reason, the most common use of TCE is to remove grease from fabricated metal parts. With wide usage for many decades, TCE has resulted in widespread contamination. The EPA has identified more than 1,400 hazardous waste sites as being the most serious in the nation. In addition to being an environmental concern, the solvent has a long list of human hazards as well. TCE is carcinogenic to humans by all routes of exposure and poses a potential human health hazard for non-cancer toxicity to the central nervous system, kidney, liver, immune system, male reproductive system, and the developing embryo/fetus.

Even with the dramatic reduction in reported TCE usage in Massachusetts, it is important to note that not reporting does not equate to not using. In addition, many in the pollution prevention assistance field consider this field to be "conquered" and have moved on to other niches to focus their efforts on the next environmental battle. In many cases, particularly among larger companies, uses and emissions of TCE have been minimized, but not eliminated, so the work continues.

Identifying a cleaning substitute for solvents is not an easy task. There are thousands of formulations to choose from, and even after selecting a potential product based on literature review, there is no guarantee that it will work. This means that after an alternative solvent (or solvents) is found, client-specific testing may be required for the change to be approved.

Replacement of TCE has been on a case-by-case basis and an all encompassing systematic, scientific approach has been generally lacking. Over the past twenty years, several tools were developed to assist in the search for safer alternatives. Rising to the top of these tools has been the CleanerSolutions online database. This system was created so that access to the testing data from the TURI Cleaning Lab could be performed easily and quickly. The end goal was to allow users to search for an alternative cleaning product based on safety and effectiveness.

Lab testing is performed in a five-step process, which includes product selection, chemistry evaluation, mechanical agitation, lab piloting, and field-testing. During each phase, the laboratory testing is tailored to the specifics of the current project. There is no one size fits all, no silver bullet. The TURI Cleaning Lab specializes in meeting the needs of the companies with which they work. In each phase, the number of viable products is reduced until only one or two remain, making selection by the requesting company easier and more likely adoptable at the end of the process.

Obstacles in making the change within a company after completing laboratory testing can stall out a project. However, getting all parties involved committed to working together to overcome the road blocks will result in a successful project. Getting the resources needed in place and all of the technical questions answered will help upper management to provide the needed approval to keep the process moving.

Remember, replacing TCE is not something that needs to be done by oneself. There are many ways to get help. This could include collaborating with local trade organizations, state agencies, federal assistance programs, vendors, and private consultants. It is also important to recognize the benefits of identifying internal experts who can champion the effort.

After more than twenty years of searching, it is evident that a quick, effective solution to a difficult problem does not exist. In the absence of the elusive TCE silver bullet, the TURI Cleaning Lab's systematic approach has been developed, refined, and utilized to aid in the search for replacements, drop-ins, or process changes. The alternatives assessment process includes a way to compare the current solvent being used with the potential solvent or process change successor based on performance and EHS criteria. Although there is no silver bullet substitution for TCE; but luckily the silver lining is the ability to adopt the TURI Cleaning Lab's assessment process. Every bad situation has some good aspect to it. For replacing TCE, the TURI Cleaning Lab and its methodology is definitely a comforting and hopeful prospect to those looking to switch to safer solvents in cleaning applications.

REFERENCES

[1] Dictionary Definition Silver Bullet. Dictionary.com. Dictionary.com, n.d. Web. 5 June 2015. <http://dictionary.reference.com/browse/ silver+ bullet>.

[2] Dictionary/Thesaurus. *The Free Dictionary*. Farlex, n.d. Web. 5 June 2015. http://www.thefreedictionary.com/silver+bullet>.

[3] Reference Idiom Silver Bullet. Using English.com. N.p., n.d. Web. 5 June 2015. http://www.usingenglish.com/reference/idioms/silver+ bullet.html#G4uD4V3DjZ1XJHQQ.99>.

[4] "Phrase Finder Silver Bullet." Phrase Finder. Phrasefinder.com, n.d. Web. 5 June 2015. <http://www.phrases.org.uk/meanings/silver-bullet.html>.

[5] Marshall, Jason P. "Road Map for Cleaning Product Selection for Pollution Prevention." Applications, Processes, and Controls. Ed. Barbara Kanegsberg and Edward Kanegsberg. 2nd ed. Boca Raton: CRC Taylor & Francis Group, 2011. 25-1-25-18. Print. Vol. 2 of *Handbook for Critical Cleaning*. 2 vols.

[6] LeBlanc, Carole A. "Search for Safer and Greener Chemical Solvents, Dr. Carole LeBlanc 2002." Diss. Erasmus U, 2002. Print.

[7] Durkee, John B., II. Management of Industrial Cleaning Technology and Processes. Amsterdam: Elesevier, 2006. Print.

[8] Kanegsberg, Barbara, and Edward Kanegsberg, eds. Cleaning Agents and Systems. 2nd ed. Boca Raton: CRC Taylor & Francis Group, 2011. Print. Vol. 1 of Handbook for Critical Cleaning. 2 vols.

[9] Draft Toxicological Profile for Trichloroethylene. Washington DC: U.S. Department of Health and Human Services, 2014. Print.

[10] OPPT Trichloroethylene (TCE) Draft Risk Assessment. Washington DC: US EPA, 2013. Print.

[11] “Basic Information.” Significant New Alternatives Policy. Environmental Protection Agency, n.d. Web. 8 June 2015. <http://www.epa.gov/ozone/snap/>.

[12] “Welcome to The MicroCare Family of Precision Cleaners.” MicroCare Precision Cleaners. MicroCare, n.d. Web. 8 June 2015. <https://www.microcare.com/faqs-detail.html?question_id=123>.

[13] Toxicological Profile for Trichloroethylene. Atlanta: Agency for Toxic Substances and Disease Registr, 2015. Toxic Substances Portal - Trichloroethylene. Web. 5 June 2015. http://www.atsdr.cdc.gov/ToxProfiles/tp.asp?id=173&tid=30>.

[14] U.S. Environmental Protection Agency. Trichloroethylene TEACH Chemical Summary. N.p.: n.p., 2007. Print. Toxicity and Exposure Assessment for Children’s Health.

[15] Gilbert, Kathleen M., and Sarah J. Blossom. Trichloroethylene: Toxicity and Health Risks. London: Humana, 2012. Print. Molecular and Intergrative Toxicology.

[16] Chiu, Weihsueh A., et al. “Human Health Effects of Trichloroethylene: Key Findings and Scientific Issues.” Environmental Health Perspect 121.3 (2013): 303-11. Print.

[17] TURI Fact Sheet Trichloroethylene. Lowell: Toxics Use Reduction Institute, 2009. Print.

[18] Marshall, Jason P., and Heidi A. Wilcox. "Drop-in Alternatives to Trichloroethylene." Process Cleaning Magazine Sept/Oct (2006): 50-57. Print.

[19] Marshall, J.P., Hands-on assistance improves already successful pollution prevention services of the..., J Clean Prod (2010), doi:10.1016/j.jclepro.2010.05.020

[20] Technical Database Services, Inc. “PARIS II A USEPA-Developed Program for Assisting the Replacement of Industrial Solvents.” TDS Numerica Factsheet. 2004. Technical Database Services, Inc. Aug. 2007 http://www.tds-tds.com/parfact.htm

[21] Kuhl, Klaus, and Lothar Lißner. CleanTool. Aug. 2007. European Union General Directorate “Enterprise”, Innovation Programme. Sept. 2007 <http://www.cleantool.org>.

[22] Rossi, Mark, and Lauren Heine. “The Green Screen for Safer Chemicals: Evaluating Flame Retardants for TV Enclosures.” Clean

Production Action. 2007. Clean Production Action. 15 Oct. 2007 http://www.cleanproduction.org/Green.Greenscreen.php

[23] "Full GreenScreen Method." The GreenScreen Method. Clean Production Action, n.d. Web. 5 June 2015. <http://www. greenscreenchemicals.org/method/full-greenscreen-method>.

[24] Marshall, Jason P. "Cleanersolutions: A Tool for Toxics Use Reduction through Solvent Substitution in Surface Cleaning Applications." Diss. U Massachusetts Lowell, 2008. Print.

[25] Marshall, Jason P. "Cleanersolutions: A Tool for Toxics Use Reduction through Solvent Substitution in Surface Cleaning Applications." Diss. U Massachusetts Lowell, 2008. Print.

[26] Kusz, JohnPaul. "Diffusion, Confusion or Illusion? Measuring the Success Rate of Alternative Cleaning Processes." CleanTech 2002 Conference Proceedings. Cleaning Resources Conference, Atlanta, GA May 2002. N.p.: Gardner Publications, 2002. N. page

[27] Nelson, William, Waste Management Resource Center (WMRC), Champaign, IL, <http://www.wmrc.uiuc.edu>. 2001

[28] "EPA's DfE Standard for Safer Products." Design for the Environment. Environmental Protection Agency, Apr. 2011. Web. 03 Mar. 2014. <http://www.epa.gov/dfe/pubs/projects/gfcp/standard-for-safer-products.pdf>.

[29] "Design for the Environment Program Master Criteria for Safer Ingredients." Design for the Environment Publications. Environmental Protection Agency, Sept. 2012. Web. 11 Mar. 2014. <http://www.epa.gov/dfe/pubs/projects/gfcp/dfe_master_criteria_safer_ingredients_v2_1.pdf>

[30] Green Seal Standard for Cleaning and Degreasing Agents. Rep. 2.1nd ed. Washington: Green Seal, 2013. Print. GS 34.

[31] "Advantages of Ecologo Certification." UL Environment. UL Environment, n.d. Web. 5 June 2015. <http://uldialogue.ul.com/ ecologo?gclid=CNWrhe266cMCFe7m7AodxSUAXw>.

[32] "Surfactant Attributes." Cleangredients. GreenBlue, n.d. Web. 5 June 2015. <[https://www.cleangredients.org/ingredient-modules/surfactants/ attributes]>.

[33] Abbott, Steven, et al. Hansen Solubility Parameters in Practice Complete with eBook, software and data. 4th ed. Digital file.

[34] Durkee, John B., II, and Jason P. Marshall. "Vapor Degreasing with Binary Azeotropes." Process Cleaning Magazine March/April (2007): 35-39. Print.

[35] “Northeast Waste Management Officials' Association.” Northeast Waste Management Officials' Association. Northeast Waste Management Officials' Association, n.d. Web. 8 June 2015. <http://www.newmoa.org>.

[36] “Environmental Sustainability Resource Center.” Environmental Sustainability Resource Center. U.S. EPA Regions 3 and 4, KPPC and the P2Rx™ Network of Regional Centers., n.d. Web. 8 June 2015. <http://esrconline.org/>.

[37] “Promoting Pollution Prevention Through Information Exchange.” Great Lakes Regional Pollution Prevention Roundtable. GLRPPR, n.d. Web. 8 June 2015. <http://www.glrppr.org/>.

[38] “About Us.” Center for Environmental Excellence. Zero Waste Network, n.d. Web. 8 June 2015. <http://www.zerowastenetwork.org/>.

[39] “About Us.” Pollution Prevention Regional Information Center. Pollution Prevention Regional Information Center, n.d. Web. 8 June 2015. <http://p2ric.org/>.

[40] “About Peaks to Praries.” Peaks to Prairies Pollution Prevention Information Center EPA Region 8. Peaks to Prairies, n.d. Web. 8 June 2015. <http://peakstoprairies.org/>.

[41] “About WSPPN.” Western Sustainability and Pollution Prevention Network. Western Sustainability and Pollution Prevention Network, n.d. Web. 8 June 2015. <http://wsppn.org/>.

[42] “Practical Solutions for Economic and Environmental Vitality.” Pacific Northwest Pollution Prevention Resource Center. PPRC, n.d. Web. 8 June 2015. <http://www.pprc.org/>.

[43] “About Center for Clean Products.” Institute for a Secure and Sustainable Environment. U of Tennessee Knoxville, n.d. Web. 8 June 2015. <http://isse.utk.edu/ccp/>.

[44] “About NYSP2I.” New York State Pollution Prevention Institute. New York State Pollution Prevention Institute, n.d. Web. 8 June 2015. <http://www.rit.edu/affiliate/nysp2i/>.

[45] “About Institute History.” Institute for Sustainability, Energy, and Environment University of Illinois at Urbana-Champaign. U OF Illinois at Urbana-Champaign, n.d. Web. 8 June 2015. <http://sustainability.illinois.edu/>.

[46] “About the Cente Who We Are.” Lowell Center for Sustainable Production, U Massachusetts Lowell, n.d. Web. 8 June 2015. <http://www.sustainableproduction.org/>.

[47] "Pollution Prevention (P2)." Pollution Prevention. Environmental Protection Agency, n.d. Web. 8 June 2015. <http:// www2.epa.gov/p2>.

[48] "National Pollution Prevention Roundtable." National Pollution Prevention Roundtable. National Pollution Prevention Roundtable, n.d. Web. 8 June 2015. <http://www.p2.org/>.

[49] "What We Do History." US Small Business Administration. US Small Business Administration, n.d. Web. 8 June 2015. <https://www. sba.gov/>.

[50] Designation of TURA Higher & Lower Hazard Substances in Massachusetts. Boston: Commonwealth of Massachusetts Executive Office of Energy and Environmental Affairs, 2015. Print.

[51] "The Toxics Release Inventory Search." Toxics Release Inventory. Environmental Protection Agency, n.d. Web. 8 June 2015. <http://www.epa.gov/enviro/facts/tri/search.html>.

About the Authors

Jason P. Marshall, ScD

Lab Director
Toxics Use Reduction Institute
University of Massachusetts Lowell
Lowell, MA, US

Dr. Marshall runs the TURI Cleaning Lab that helps to identify and validate the performance of industrial cleaning solvents. Throughout the past two decades, the concept of toxics use reduction has been paramount to the research efforts. As part of the recent changes to the research conducted at the TURI Lab, more effort has been made to expand the research direction of the TURI Lab. Originally the lab's focus was pollution prevention assistance research to companies looking for safer work practices. The expansion has moved up the supply chain to work directly with the manufacturers of the cleaning products to assist in the development of safer cleaning formulations.

During his time at TURI, he has developed a program at UML to provide hands on training for graduate and undergraduate students in a research laboratory. The program has expanded rapidly from just a few students to more than twenty in less than five years. While in the program students are mentored and trained on practical lab skills including all phases of research.

Heidi A. Wilcox, MS
Field Specialist
Toxics Use Reduction Institute
University of Massachusetts Lowell
Lowell, MA, US

Ms. Wilcox is the field specialist and scientist at the TURI Cleaning Lab. She works on integrating lab work with field work in different types of facilities. Her areas of expertise include process engineering, one-on-one technical assistance with clients, piloting in the field, vendor interaction, and identifying safer alternatives in the industrial cleaning industry. Heidi has worked with many companies on performance testing, R&D, consulting and training and is committed to finding less toxic solvents and cleaning processes that work, are economical and safer for worker and environment. She earned a Bachelor's of Science degree in Microbiology from UMass Amherst and a Master of Science in Environmental Studies from UMass Lowell. Currently, she is a finishing her doctoral degree in the field of Cleaner Production and Toxics Use Reduction at UMass Lowell.

Index

A

accelerated eutrophication, 6
acid rain formation, 8
adhesives, 11, 27
air sparging, 30
aircraft manufacturing, 151
alkaline aqueous, 68, 69
alternatives assessment, 2, 153, 159
aluminum, 44, 94, 131, 132, 136, 137, 138, 139
American society for testing and materials (ASTM), 28, 89
analytical assessment, 62
animal studies, 85
animal testing, 87
aqueous cleaners, 6, 9, 41
atmospheric ozone depletion, 8

B

bioaccumulation, 88
biological degradation, 23
boiling point, 15, 17, 94
buffing compounds, 12
byproduct, 26

C

cancer, 22, 23, 24, 81, 85, 149, 158
carbon tetrachloride, 7
carbonates, 6
carcinogenic, 3, 22, 81, 85, 86, 157
carcinogenicity, 8, 42
central nervous system, 8, 21, 22, 24, 158
ceramic, 30, 131
chelating agents, 90
chemical action, 59, 60, 154
chemical data report (CDR), 20
chemical feedstock, 7
chemical hazard, 78
chemical stabilizers, 17
chemical waste sites, 20
chloral hydrate, 21, 24
Clean air act (CAA), 16
Cleangredients, 164
cleaning, vii, xi, 1, 2, 3, 4, 5, 6, 7, 8, 9, 11, 12, 14, 15, 16, 17, 19, 25, 26, 27, 28, 29, 30, 31, 33, 35, 36, 37, 40, 41, 42, 44, 54, 55, 57, 59, 60, 61, 62, 63, 64, 65, 66, 72, 73, 74, 75, 77, 78, 79, 85, 86, 88, 89, 90, 91, 92, 93, 95, 97, 98, 100, 107, 120,123, 124, 126, 127, 128, 129, 130, 131, 132, 135, 136, 137, 138, 139, 140, 144, 145, 149, 154, 155, 157, 158, 159, 167, 168
cleaning equipment, 3

Cleaning Laboratory at UMass Lowell, 2
cleaning parameters, 59
client based testing, 31
Comprehensive Environmental Response Compensation and Liability act (CERCLA), 20, 90
contaminants, 3, 6, 12, 13, 40, 42, 89, 130
contamination, 5, 8, 16, 19, 86, 125, 157
cutting fluids, 12, 44

D

degreasing, 8, 9, 13, 15, 17, 19, 21, 23, 26, 67, 69, 73, 86, 89, 93, 95, 108, 120, 130, 131, 132, 133, 135, 137
Department of Defense (DOD), 143
detergency, 9
detergents, 6, 9, 79
Dichloromethane (CH2CL2), 8
dip tank, 30
dipolar intermolecular force, 16
dispersion forces, 16

E

EASTEC, 100
electronics, 3, 30, 107
Emergency Planning and Community Right to know act (EPCRA), 20, 38, 90
emulsifying agent, 6
engineering controls, 25, 140
Environmental Health and Safety (EH&S), 39, 77, 98, 112, 114, 118, 119, 120
Environmental Protection agency (EPA), viii, xii, 13, 16, 19, 20, 22, 24, 78, 79, 81, 85, 90, 92, 101, 102, 103, 104, 108, 135, 142, 143, 144, 145, 148, 149, 153, 155, 157, 162, 163, 164
Environmental sustainability resource center (ESRC), 102
EPA draft risk assessment, 13
EU classification, labeling and packaging, EU CLP, 85
exposure potential, 148
exposure profile, 148
exposures, 3, 22, 149

F

fabricated metal parts, 11, 157
fatty acids, 6
Federal drug administration (FDA), 4, 31, 65, 115
field testing, 29, 57
flammability, 8, 42, 77
flammable, 3, 41, 78
flash point, 15
formulators, 3, 89, 92
fugitive emissions, 23

G

geometry, 60, 62, 121
glass, 30, 79, 129
global consumption, 20
global warming, 3, 7, 8, 42
Globally Harmonized system, GHS, 79, 81, 82, 83, 84
grease, 6, 9, 11, 20, 30, 44, 157
Great Lakes Regional Pollution Prevention Roundtable (GLRPPR), 102, 164
Green Screen, vii, 39, 40, 162
groundwater, 20, 23, 148

H

halogenated compounds, 7
hands on training, 167
hazardous, xii, 3, 7, 19, 20, 34, 36, 39, 41, 42, 69, 74, 101, 112, 116, 117, 118, 119, 157
hazardous chemical, 3
hazardous waste sites, 19, 157
health and safety, 3, 33, 35, 39, 44, 67, 75, 94, 107
health effects, 8, 19, 24
human carcinogen, 24, 81, 148
human health hazard, 22, 148, 149, 157

Hydrochlorofluorocarbon (HCFC), 68, 69, 73
Hydrofluroethers (HFE), 7, 68, 69, 70, 71, 72, 73, 93, 132, 133
hydrogen bonds, 16

I

immersion, 8, 15, 30, 130, 132, 137, 138
immersion cleaning, 15, 130, 132, 137, 138
immune system, 22, 158
industrial parts cleaning, 3
Institute for Research & Technical Assistance (IRTA), 107
Integrated Risk Information System (IRIS), 22
Integrated solvent substitution database (ISSDS), 34

K

KB Value, 15

L

lab piloting, 57, 158
laboratory, xi, 1, 2, 44, 54, 57, 60, 62, 65, 93, 95, 107, 123, 124, 129, 130, 133, 138, 143, 144, 145, 153, 154, 155, 158, 167
latent heat, 15
life cycle assessment, 90
low solubility, 7
Lowell Center for Sustainable Production, 107, 165
lubricants, 12, 44

M

manual scraping, 30
manual wiping, 30
marble, 30
Massachusetts Office of Technical Assistance, xii, 144, 145
Maximum contaminant level (MCL), 20
mechanical energy, 128, 138
metal degreasing, 23
metal working, 12, 36, 99, 131, 136, 145, 151
methyl ethyl ketone, 8
methylene chloride, 8, 15, 16, 119
Minimum risk level (MRL), 20

N

Narragansett Bay Commission (NBC), 144, 145
National Pollution Prevention Roundtable, 108, 165
National priorities list (NPL), 19
neurotoxicity, 8
nickel, 30, 125
North American Industrial Classification Systems codes (NAICS), 13
Northeast waste management officials association (NEWMOA), 101

O

occupational limits, 22
oil, 9, 86, 89, 94, 124, 125, 131, 132, 136, 137
oleoresins, 6
on-site assistance, 66, 145
optimum efficiency, 59
organic, 3, 6, 7, 8, 9, 11, 16, 22, 42, 85, 87, 89, 157
organic contaminants, 6, 89
ozone depletion, 3, 7, 8, 42

P

P2, 98, 99, 100, 101, 102, 108, 152, 153, 154, 155, 165
Pacific Northwest Pollution Prevention Resource Center (PPRC), 104, 105, 164

paint removers, 11
painting, 5, 10, 30
part shape, 62, 154
pepper spray, 11
Perchloroethylene (PCE), 8, 119, 135
petroleum, 8, 77
pH, 38, 41, 42, 68, 72, 73, 86, 90, 91, 126, 127, 132
phosgene, 20, 24
phosphate builders, 6
phosphorus, 6, 85, 87
pilot testing, 64, 75, 155
plastic, 30, 131
pollution prevention, xi, 25, 63, 97, 98, 101, 102, 103, 105, 108, 116, 120, 140, 147, 148, 158, 162, 167
Pollution Prevention Options Analysis System (P2OASys), vii, viii, 3, 34, 36, 37, 39, 40, 43, 67, 74, 75, 129, 133
Pollution Prevention Regional Information Center (P2RIC), 103, 105, 164
polymers of isoprene, 6
probable human carcinogen, 24
process change, 2, 36, 115, 116, 118, 157, 159
product classifications, 73
product selection, 57, 136, 153, 158

R

reproductive effects, 23
reproductive toxicity, 8, 86
resource use, 88
Rhode Island Department of environmental management (RI DEM), 143, 144, 145
RI OSHA, 144, 145
risk, 3, 13, 16, 22, 23, 36, 41, 77, 78, 80, 81, 82, 83, 85, 90, 116, 148, 149
risk assessment, 13, 90, 148, 149

S

screening, 3, 39, 42, 44, 54, 67, 68, 69, 73, 74, 75, 93, 126, 131, 134, 148
Significant new use rule (SNURs), 149
silver bullet, xi, xii, 1, 2, 4, 7, 8, 9, 55, 57, 66, 73, 95, 158, 159
silver lining, xi, 2, 159
smog, 7, 77
Snap approved (Significant New Alternatives Policy), 16, 162
soaps, 6
Society of manufacturing engineers (SME), 100
soil management, 5
soils, 3, 6, 9, 11, 12, 17, 29, 58, 60, 65, 73, 86, 89, 94, 118, 120, 126, 128, 129, 130, 132, 133, 136, 138, 143, 154, 155, 157
solvating power, 3, 6, 8
solvent alternative guide (SAGE), 34
solvent substitution, xi, 42, 65, 123
solvents, xi, xii, 2, 3, 7, 8, 9, 14, 16, 17, 22, 25, 29, 30, 34, 35, 41, 42, 44, 64, 67, 68, 73, 77, 90, 93, 94, 95, 108, 112, 117, 119, 120, 121, 124, 130, 131, 132, 134, 135, 155, 158, 159, 167, 168
spot removers, 11
stainless steel, 30, 125, 137, 138
stakeholders, 78, 79, 106, 149
standardized testing procedures, 3
steel, 30, 86, 89, 125, 131, 136, 137, 138, 139
stratospheric ozone, 16
substitution, xi, 25, 36, 42, 44, 57, 65, 79, 86, 123, 140, 157, 159
substrates, 3, 29, 40, 73, 118, 120, 138
Superfund Amendments and reauthorization act (SARA), 20, 90
supply chain, 30, 116, 117, 120, 144, 167
surface tension, 9, 15, 121, 136
surface type, 131
surfactants, 9, 89, 90, 164
synthesized compounds, 7
synthetic soaps, 6

T

TACT, 59, 61, 62, 128, 154
TCE silver bullet, xi, 2, 4, 95, 157, 159

technical assistance, 3, 43, 97, 99, 101, 102, 103, 104, 107, 112, 114, 116, 140, 143, 168
technical assistance providers, 43, 101, 102, 114
temperature cleaning, 59
terpenes, 6, 90
testing trials, 3
The New York State Pollution Prevention Institute (NYSP2I), 106, 164
thermodynamics, 92
titanium, 30
toxic, 2, 3, 7, 21, 24, 80, 81, 83, 85, 87, 93, 104, 106, 107, 113, 116, 135, 147, 149, 155, 168
Toxics release inventory (TRI), 3, 147, 149
Toxics substance control act (TSCA), 85, 90, 148
toxics use reduction, 2, 25, 36, 167
Toxics use reduction Act (TURA), 2, 26, 107, 113, 165
Toxics Use Reduction institute (TURI), vii, xi, 2, 4, 12, 13, 16, 27, 29, 33, 40, 41, 42, 54, 55, 57, 58, 60, 61, 63, 64, 65, 66, 67, 68, 73, 74, 75, 93, 97, 107, 123, 124, 126, 135, 140, 143, 144, 145, 147, 148, 149, 151, 153, 154, 155, 157, 158, 159, 162, 167, 168
Trichloroethylene (TCE), v, vii, viii, ix, xi, xii, 1, 2, 3, 4, 8, 9, 11, 12, 13, 14, 15, 16, 17, 19, 20, 21, 22, 23, 24, 25, 26, 27, 28, 33, 55, 57, 64, 66, 67, 68, 69, 72, 74, 75, 77, 86, 93, 94, 95, 97, 98, 100, 109, 112, 113, 115, 117, 118, 119, 121, 123, 124, 125, 126, 129, 131, 133, 134, 135, 140, 142, 143, 144, 145, 147, 148, 149, 150, 151, 152, 153, 155, 157, 158, 159, 161, 162
TURI Cleaning Lab, vii, 2, 4, 12, 13, 16, 27, 29, 33, 40, 41, 42, 54, 55, 57, 58, 60, 61, 63, 64, 65, 66, 67, 68, 73, 74, 75, 93, 123, 124, 126, 135, 140, 143, 144, 145, 147, 148, 149, 151, 153, 154, 155, 158, 159, 167, 168

U

ultrasonic cleaning, 137
ultrasonic transducers, 13
ultrasonics, 44, 120, 128
UMass Lowell, 2, 168
United Laboratories ECOLOGO, 78, 88, 92
universal solvent, 6, 8
University of Tennessee Center for Cleaner Production, CCP, 105
US Small Business Administration (SBA), 108, 165

V

vapor degreasing, 8, 9, 13, 15, 17, 67, 69, 73, 93, 95, 108, 120, 130, 131, 132, 133, 135, 137
vapor phase, 8, 15, 24
vapor phase cleaning, 8
vapor pressure, 7, 8, 15, 23, 87
Volatile Organic Compounds (VOC), 3, 7, 38, 41, 42, 68, 72, 73, 87, 91, 94, 126, 127, 130, 132

W

water, xi, 6, 8, 9, 10, 11, 15, 20, 22, 23, 59, 60, 88, 94, 95, 106, 115, 118, 120, 121, 127, 131, 136, 144, 148
water softeners, 6
waxes, 137
welding, 5, 10
Western Sustainability and Pollution Prevention Network (WSPPN), 103, 105, 164

Z

Zero Waste Network, 102, 104, 164